Andreas Rüegg

**Artificially structured strongly correlated electron systems**

Andreas Rüegg

# Artificially structured strongly correlated electron systems

## Model calculations of physical properties

Südwestdeutscher Verlag für Hochschulschriften

**Impressum/Imprint (nur für Deutschland/ only for Germany)**
Bibliografische Information der Deutschen Nationalbibliothek: Die Deutsche Nationalbibliothek verzeichnet diese Publikation in der Deutschen Nationalbibliografie; detaillierte bibliografische Daten sind im Internet über http://dnb.d-nb.de abrufbar.
 Alle in diesem Buch genannten Marken und Produktnamen unterliegen warenzeichen-, marken- oder patentrechtlichem Schutz bzw. sind Warenzeichen oder eingetragene Warenzeichen der jeweiligen Inhaber. Die Wiedergabe von Marken, Produktnamen, Gebrauchsnamen, Handelsnamen, Warenbezeichnungen u.s.w. in diesem Werk berechtigt auch ohne besondere Kennzeichnung nicht zu der Annahme, dass solche Namen im Sinne der Warenzeichen- und Markenschutzgesetzgebung als frei zu betrachten wären und daher von jedermann benutzt werden dürften.

Verlag: Südwestdeutscher Verlag für Hochschulschriften Aktiengesellschaft & Co. KG
Dudweiler Landstr. 99, 66123 Saarbrücken, Deutschland
Telefon +49 681 37 20 271-1, Telefax +49 681 37 20 271-0
Email: info@svh-verlag.de
Zugl.: Zürich, ETH Zürich, Diss., 2009

Herstellung in Deutschland:
Schaltungsdienst Lange o.H.G., Berlin
Books on Demand GmbH, Norderstedt
Reha GmbH, Saarbrücken
Amazon Distribution GmbH, Leipzig
**ISBN: 978-3-8381-1494-1**

**Imprint (only for USA, GB)**
Bibliographic information published by the Deutsche Nationalbibliothek: The Deutsche Nationalbibliothek lists this publication in the Deutsche Nationalbibliografie; detailed bibliographic data are available in the Internet at http://dnb.d-nb.de.
 Any brand names and product names mentioned in this book are subject to trademark, brand or patent protection and are trademarks or registered trademarks of their respective holders. The use of brand names, product names, common names, trade names, product descriptions etc. even without a particular marking in this works is in no way to be construed to mean that such names may be regarded as unrestricted in respect of trademark and brand protection legislation and could thus be used by anyone.

Publisher: Südwestdeutscher Verlag für Hochschulschriften Aktiengesellschaft & Co. KG
Dudweiler Landstr. 99, 66123 Saarbrücken, Germany
Phone +49 681 37 20 271-1, Fax +49 681 37 20 271-0
Email: info@svh-verlag.de

Printed in the U.S.A.
Printed in the U.K. by (see last page)
**ISBN: 978-3-8381-1494-1**

Copyright © 2010 by the author and Südwestdeutscher Verlag für Hochschulschriften Aktiengesellschaft & Co. KG and licensors
All rights reserved. Saarbrücken 2010

# Abstract

The amazing experimental progress in fabrication processes of layered structures has led to a fascinating new research field: artificial structures of complex materials with strongly correlated electron systems. The bright perspective for intriguing new physics with potential for applications is based on two important facts. First, artificial structures offer new options to control the material properties and it is possible to reach regimes which are not accessible in bulk. Second, materials with strongly correlated electron systems show a rich variety of electronic and magnetic states. Their phase diagrams are considerably richer compared to semiconductors and standard metals. It is therefore likely that remarkable new phenomena are observed in artificial structures with strongly correlated electron systems. This new field is still in its infancy and many fundamental as well as more applied questions have to be investigated.

This thesis deals with the theory of strong correlation physics in heterostructures containing Mott and band insulators. Such systems are already realized in experiments using different transition metal oxides. Because these materials contain chemical bonding with a substantial ionic character, electrostatic considerations play an essential role in explaining the physical properties of the layered structures. Notably, the ionic charges of a layered structure may induce a large dipole field. If the latter is compensated by a rearrangement of electronic charges a conducting interface is formed between the two materials. Our goal is to characterize this metallic phase.

In order to address the consequences of strong local correlations in the metallic state we use an extension of the slave-boson mean-field approximation introduced by Kotliar and Ruckenstein. A part of the thesis is therefore devoted to technical aspects related to this approach. Specifically, we develop a formulation which allows to include incoherent atom-like features in the single-particle spectral weight and we discuss in detail the generalization of the original formulation to the inhomogeneous situation with long-range Coulomb interaction.

As a generic model to investigate the physical properties of layered band insulator/ Mott insulator structures we use a single-orbital Hubbard model extended to heterostructures. We investigate the interplay between the electrostatic problem and the effect of a strong local repulsion between the electrons. Within the slave-boson mean-field approximation we find that the quasiparticles responsible for metallic behavior are subject to strong renormalization effects. In particular, itinerant and localized degrees of freedom hybridize in the interface region. This kind of hybridization is reminiscent of heavy fermion physics in homogeneous systems.

The characteristic features of the interface are further studied by investigating the low-temperature quasiparticle transport of the layered structure from a semiclassical perspective. We focus on two quantities: the low-frequency optical conductivity and the thermoelectrical power. For both quantities, the contribution from the interface is of particular importance. In the limit of low temperatures the thermoelectrical power is enhanced if the effective hybridization between itinerant and localized degrees of freedom is small. It is especially large in situations with a sharp change in the spatial distribution of the electronic charge at the interface and in systems with a thick layer of the Mott insulator. This low-temperature behavior is in contrast to the behavior in the high temperatures limit which we describe by a generalization of Heikes formula to inhomogeneous systems.

The field of nanoscale structured strongly correlated electron systems is rapidly growing and we hope that this thesis can provide useful inputs for both experiments and further theoretical studies. In particular, our results show that quantum mechanical coherence over several atomic monolayers in the vicinity of the interface leads to interesting and unusual phenomena. Surely, many more important aspects will be discussed in the future. They include various short- and long-range ordered phases in spin and orbital degrees of freedom, superconductivity as well as problems related to the physics of multiferroics and possible applications as spintronic devices.

# Contents

| | |
|---|---|
| Abstract | i |
| Contents | iii |
| List of Figures | vii |
| Acronyms | ix |

**1 Introduction**    1
    1.1 From Bloch-Sommerfeld to strongly correlated artificially structured electron systems . . . . . . . . . . . . . . . . . . . . . . . . 1
    1.2 Outline . . . . . . . . . . . . . . . . . . . . . . . . . . . . . . . 3

**2 A few basic concepts and a model for correlated heterostructures**    5
    2.1 Introduction . . . . . . . . . . . . . . . . . . . . . . . . . . . . 5
       2.1.1 Experimental realizations . . . . . . . . . . . . . . . . . 6
       2.1.2 Classification . . . . . . . . . . . . . . . . . . . . . . . . 7
    2.2 Reconstruction and the polar catastrophe . . . . . . . . . . . . 9
       2.2.1 Macroscopic electrostatic considerations . . . . . . . . . 9
       2.2.2 Relaxation and interface reconstruction . . . . . . . . . 12
    2.3 Theoretical approaches . . . . . . . . . . . . . . . . . . . . . . 13
    2.4 Model of a homometallic heterostructure . . . . . . . . . . . . 13
       2.4.1 Hubbard heterostructure . . . . . . . . . . . . . . . . . 14
       2.4.2 Thomas-Fermi screening and charge leakage . . . . . . . 16

# CONTENTS

## 3 Slave-particle methods for strong correlation physics    23
- 3.1 A brief overview . . . . . . . . . . . . . . . . . . . . . . . . . 23
  - 3.1.1 Schwinger-Wigner representation . . . . . . . . . . . . . 24
  - 3.1.2 Barnes-Coleman representation . . . . . . . . . . . . . . 24
  - 3.1.3 Kotliar-Ruckenstein representation . . . . . . . . . . . . 25
- 3.2 Four boson mean-field approximation . . . . . . . . . . . . . . . 26
  - 3.2.1 Formalism for the single orbital Hubbard model . . . . . . 26
  - 3.2.2 The Mott transition in the two-band Hubbard model . . . 31
- 3.3 A slave-spin theory for the particle-hole symmetric Hubbard model 33
  - 3.3.1 Model and method . . . . . . . . . . . . . . . . . . . . . 34
  - 3.3.2 Mean-field theory . . . . . . . . . . . . . . . . . . . . . . 36
  - 3.3.3 Brinkman-Rice transition . . . . . . . . . . . . . . . . . . 38
  - 3.3.4 Schwinger bosons and pseudospins . . . . . . . . . . . . . 39
  - 3.3.5 Role of fluctuations in three dimensions . . . . . . . . . . 40
- 3.4 The SBA approximation and DMFT . . . . . . . . . . . . . . . . 45
  - 3.4.1 The dynamical mean-field approach . . . . . . . . . . . . 46
  - 3.4.2 SBA impurity solver . . . . . . . . . . . . . . . . . . . . 48

## 4 Strongly correlated electrons in a Hubbard heterostructure    51
- 4.1 Introdcution . . . . . . . . . . . . . . . . . . . . . . . . . . . . 51
- 4.2 Quasiparticle description . . . . . . . . . . . . . . . . . . . . . . 53
- 4.3 Slave-boson mean-field treatment . . . . . . . . . . . . . . . . . 55
  - 4.3.1 Ambiguity of Schrödinger-Poisson-Gutzwiller . . . . . . . 56
  - 4.3.2 Effective one-dimensional Schrödinger equation . . . . . . 57
  - 4.3.3 Longe range Coulomb interaction . . . . . . . . . . . . . 58
  - 4.3.4 Free energy and self-consistency . . . . . . . . . . . . . . 59
  - 4.3.5 Numerical scheme . . . . . . . . . . . . . . . . . . . . . 60
- 4.4 Results and discussion . . . . . . . . . . . . . . . . . . . . . . . 61
  - 4.4.1 Ground-state properties . . . . . . . . . . . . . . . . . . 61
  - 4.4.2 Metallic interfaces in the Mott regime . . . . . . . . . . . 65
  - 4.4.3 Quasiparticle properties . . . . . . . . . . . . . . . . . . 68
- 4.5 Interfacial heavy-fermion scenario . . . . . . . . . . . . . . . . . 74
- 4.6 Comparison to related approaches . . . . . . . . . . . . . . . . . 76
  - 4.6.1 Comparison to two-site dynamical mean-field theory (DMFT) 76

|  | 4.6.2 | Comparison to the "pseudocanonical" Gutzwiller approximation (GA) | 78 |
|---|---|---|---|
| 4.7 | | Conclusions | 81 |

## 5 Optical conductivity and thermoelectricity in correlated superlattices — 83

- 5.1 Introduction . . . 83
- 5.2 Drude weight and optical conductivity . . . 85
  - 5.2.1 Linear response . . . 86
  - 5.2.2 Twisted boundary conditions . . . 87
  - 5.2.3 Quasi-particle contribution . . . 87
  - 5.2.4 Results . . . 90
- 5.3 Thermoelectricity . . . 92
  - 5.3.1 Generalized transport coefficients . . . 93
  - 5.3.2 Estimation of the relaxation time . . . 96
  - 5.3.3 DC conductivities . . . 99
  - 5.3.4 Thermopower . . . 99
  - 5.3.5 Dependence of the thermopower on model parameters . . . 101
  - 5.3.6 Quantum oscillations in the thermopower . . . 104
  - 5.3.7 Thermoelectrical figure of merit and power factor . . . 104
  - 5.3.8 Thermopower in the atomic limit . . . 105
- 5.4 Conclusions . . . 107

## A for chapter 3 — 109
- A.1 Pseudo-spin wave analysis . . . 109
  - A.1.1 Canonically transformed Hamiltonian . . . 109
  - A.1.2 Effective Hamiltonian for excitations . . . 109
  - A.1.3 Bogoliubov transformation . . . 111
  - A.1.4 Matrix elements for the spectral density . . . 112

## B for chapter 5 — 115
- B.1 Phenomenological thermoelectricity . . . 115
  - B.1.1 Entropy production . . . 116
  - B.1.2 Temperature distribution . . . 116
  - B.1.3 Solution of transport equations . . . 117

  B.2 Kubo formalism for transport coefficients . . . . . . . . . . . . . . 119
    B.2.1 Linear dynamical laws . . . . . . . . . . . . . . . . . . . . 119
    B.2.2 Quasiparticle transport . . . . . . . . . . . . . . . . . . . . 120
    B.2.3 Approach from the atomic limit . . . . . . . . . . . . . . . 122

**Bibliography**    **125**

**Acknowledgments**    **139**

# List of Figures

2.1 Ideal $ABO_3$ structure . . . . . . . . . . . . . . . . . . . . . . . 6
2.2 Classification of (001) $ABO_3/A'B'O_3$ interfaces . . . . . . . . . . 8
2.3 Schematic electrostatic diagram for a polar compound . . . . . . . 10
2.4 Sketch of the Hubbard heterostructure . . . . . . . . . . . . . . . 15
2.5 Density distribution in different approximations . . . . . . . . . . 21

3.1 Slave-boson fields for the two-band Hubbard model . . . . . . . . 32
3.2 Quasiparticle weight in the two-band Hubbard model . . . . . . . 32
3.3 Pseudo-spin excitation spectrum and effective mass . . . . . . . . 41
3.4 Single-particle spectral density in the slave-spin method . . . . . . 43
3.5 Fluctuation regime in the slave-spin method . . . . . . . . . . . . 45

4.1 Algorithm used for the Hubbard heterostructure . . . . . . . . . . 56
4.2 Overview of the interaction dependence of the mean fields . . . . 62
4.3 Non-monotonic dependence of the density distribution . . . . . . 64
4.4 Square mean distance of the electronic charge from the center . . 65
4.5 Coherent part of the layer resolved spectral density . . . . . . . . 67
4.6 Coherent particle density . . . . . . . . . . . . . . . . . . . . . . 68
4.7 Quasiparticle weight of the partially filled subbands . . . . . . . . 70
4.8 Dependence of the quasiparticle weight on the number of ion layers 71
4.9 Quasiparticle dispersion of the partially filled subbands . . . . . . 72
4.10 Transverse wave functions . . . . . . . . . . . . . . . . . . . . . . 73
4.11 Effective potential . . . . . . . . . . . . . . . . . . . . . . . . . . 74
4.12 Comparison of the coherent charge density with the two-site DMFT 77
4.13 Comparison with the charge density and the hopping renormalization factor in the Gutzwiller approximation . . . . . . . . . . . 78

## LIST OF FIGURES

| | | |
|---|---|---:|
| 4.14 | Comparison with the dispersion in the Gutzwiller approximation | 80 |
| 5.1 | Setup considered for the thermoelectrical transport | 84 |
| 5.2 | Drude weight and optical conductivity | 91 |
| 5.3 | Drude weight for different superlattices and the two-dimensional response | 92 |
| 5.4 | Characteristic quantities of the quasiparticle dispersion | 95 |
| 5.5 | The scattering rate as function of the impurity strength | 98 |
| 5.6 | Seebeck coefficient as function of $U_r$ | 102 |
| 5.7 | Seebeck coefficient as function of $E_C$ | 103 |
| 5.8 | Quantum oscillations in the thermopower | 105 |
| B.1 | Schematics of a thermoelectrical cooling device | 117 |

# Acronyms

**BCS**  Bardeen-Cooper-Schrieffer
**BI**  band insulator
**DMFT**  dynamical mean-field theory
**DFT**  density functional theory
**DOS**  density of states
**GA**  Gutzwiller approximation
**IF**  interface region
**LDA**  local density approximation
**LTO**  $LaTiO_3$
**LAO**  $LaAlO_3$
**MI**  Mott insulator
**OSMT**  orbital-selective Mott transition
**PAM**  periodic Anderson model
**RMFT**  renormalized mean-field theory
**SBA**  Kotliar and Ruckenstein slave-boson mean-field approximation
**SCE**  strongly correlated electron
**STO**  $SrTiO_3$

# Chapter 1

# Introduction

## 1.1 From Bloch-Sommerfeld to strongly correlated artificially structured electron systems

Electronic properties of solid-state systems are in many situations well understood in terms of the quantum mechanical Bloch-Sommerfeld description of noninteracting fermions which are subject to a periodic potential given by the details of the crystallographic structure. The quantum mechanical character of the electrons leads to the formation of energy bands which are filled according to the Pauli principle. Materials with an odd number of electrons per unit cell have partially filled bands and therefore show metallic behavior. On the other hand, semiconductors and band-insulators have completely filled bands with an energy gap between the highest filled and the lowest empty band. Electrical transport is therefore thermally activated and suppressed at low temperatures (compared to the gap). Starting from the single-particle Bloch-states, also the perturbative interaction between the electrons, the coupling of the electrons to the lattice degrees of freedom and the influence of defects and impurities can be studied. The amazing fact that electrons in metals can be considered as an accumulation of weakly interacting fermions allows for the phenomenological description in terms of Landau's Fermi liquid theory [1, 2]. It is fundamentally based on the concept of screening and on the principle of adiabatic continuity. Screening transforms the long-range Coulomb interaction between the electrons into a

## 1.1 From Bloch-Sommerfeld to strongly correlated artificially structured electron systems

short-range interaction which removes divergences appearing in a perturbative treatment of the electron-electron interaction [1]. On the other hand, adiabatic continuity states that the excitations of the interacting electron system are adiabatically connected to the excitations of a non-interacting system [3]. Nowadays, the band theory is formulated within the density functional theory (DFT) [4, 5] which provides a rigorous framework to study the electronic structure of solids and to find the "best" single-particle states, for example within the local density approximation (LDA).

On the other hand, there are numerous examples where the electronic properties are seemingly incompatible with (or at least not easily explained within) the above *standard model* of solids. The loosely defined terminology strongly correlated electron (SCE) systems has been introduced to collectively refer to condensed matter systems where the correlations between the electrons can lead to a breakdown of Landau's Fermi liquid theory. Prominent examples are so-called Mott insulators [6] – a class of materials that are expected to be metallic under conventional band theory but in fact are not due to strong electron-electron interaction. Usually, these materials are magnetic insulators at low temperatures. More general, in SCE systems, metallicity is weakened or even lost completely. The importance of correlations for the breakdown of band theory and for the insulating character of magnetic insulators has been known for several decades [7]. But it was the discovery of high-temperature superconductivity in layered cuprates [8] and the observation of the fractional quantum Hall effect in ultra clean semiconductor heterostructures [9, 10] which has let to an enormous amount of work investigating the role of strong electron correlations in attempts to explain the extraordinary properties of these systems.

Transition metal oxides are a particularly interesting class of materials that are expected to have SCE systems. In many cases, they have a rich and complex phase diagram including metal-insulator transitions, ordered phases in spin and orbital degrees of freedom, as well as various exotic liquid phases [11, 12]. The electronic and magnetic phases show a high sensitivity to external conditions such as temperature, pressure, doping and application of electric or magnetic fields. The broad spectrum of intrinsic functionalities, including high-$T_c$ and unconventional superconductivity, colossal magnetoresistance and multiferroic behavior, have also brought attention to these materials as promising candidates for applications (oxide-electronics) and have motivated people to engineer artificial nanoscale structures based on complex oxides.

Indeed, the amazing experimental progress in fabrication processes of layered structures has led to a fascinating new research field that investigates the physical properties of artificially structured SCE systems. The bright perspective for intriguing new physics with potential for applications is based on two important facts. First, the experience with semiconductors demonstrates that artificial structures allow for a better control of the material properties, for example by the electric-field-effect or by the principle of modulation doping. Furthermore, it is possible to reach regimes which are not accessible in bulk. Second, as argued above, materials with SCE systems show a rich variety of electronic and magnetic states. Their phase diagrams are considerably richer compared to semiconductors and standard metals. These class of materials is therefore likely to show remarkable new phenomena in artificial structures where the conditions change locally. Taking advantage of new mechanisms at interfaces of correlated electron systems may allow to design junctions for potential device applications and the richer properties of the SCE systems may lead to novel functionalities.

## 1.2 Outline

The thesis at hand deals with strongly correlated electron systems in artificially layered structures composed of a Mott insulator (MI) and a band insulator (BI). The outline is as follows. In Chap. 2 we discuss basic concepts related to the heterostructures built on compounds belonging to the family of perovskites. We discuss the scenario of the electronic compensation mechanism at a polar interface. Furthermore, we introduce in this chapter a generalized Hubbard model describing a Mott insulator geometrically confined in a heterostructure. The electronic properties of this model are discussed in Chap. 4 and aspects of the transport in Chap. 5. Unusual renormalization phenomena of the electronic states in the vicinity of the interface lead to features in the electronic structure which are reminiscent of heavy-fermion physics. In Chap. 5 we suggest that the low-temperature thermoelectrical effect is affected by these many-body phenomena. Our description of correlation effects in the Hubbard heterostructure is, to a large extend, based on the slave-boson approach of Kotliar and Ruckenstein.

## 1.2 Outline

We therefore have included Chap. 3 which deals with technical aspects related to this particular approach.

# Chapter 2

# A few basic concepts and a model for correlated heterostructures

> *It has recently become possible to fabricate heterostructures of complex oxides with atomically smooth interfaces and several experiments suggest a few basic concepts of particular importance for this class of systems. Here, we discuss the polar discontinuity scenario which can lead to unique reconstruction mechanisms at interfaces triggered by electrostatic boundary conditions. In order to theoretically study aspects of correlated electron physics in confined dimensions, we introduce an extended Hubbard model and discuss the role of screening by means of a generalized Thomas-Fermi approach. The investigation of this model is extended in the subsequent Chaps. 4 and 5 of this thesis to include electronic and transport properties.*

## 2.1 Introduction

Strongly correlated electron (SCE) systems are potential candidates for the next generation functional materials for various purposes and there are by now a variety of experimentally realized digital heterostructures based on SCE compounds. In particular, nanoscale structures of complex oxide materials have attracted attention during the last few years. In numerous cases, materials with a crystallo-

## 2.1 Introduction

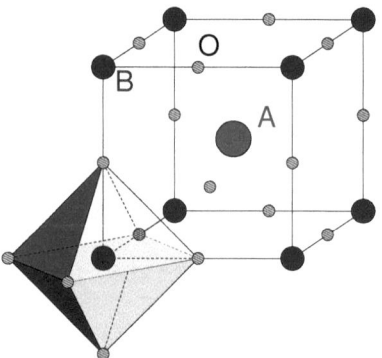

Figure 2.1: Schematics of the ideal ABO$_3$ perovskite structure.

graphic structure of the (cubic) ABO$_3$ perovskite-type are used as building blocks for lattice-matched heterostructures and to be specific, we focus here and also in the subsequent Chaps. 4 and 5 on this class of systems. The ideal cubic structure is schematically shown in Fig. 2.1. Taking oxygen to have a formal valance of O$^{2-}$, the A and B cations can take on values A$^{4+}$B$^{2+}$, A$^{3+}$B$^{3+}$, A$^{2+}$B$^{4+}$ or A$^{5+}$B$^{1+}$. We are interested in situations where the B-type cation is a metal such as Al or a transition metal such as Ti, V, Mn, or Cu. The A-cation is usually a bigger atom such as Sr, Y or La. Because of the relative simple atomic structure, the near lattice match between a variety of different bulk systems as well as the fact that in particular SrTiO$_3$ (STO) is a widely used and well controlled substrate make this class of solid-solid interfaces an interesting playground and up to date many examples can be found in the literature.

### 2.1.1 Experimental realizations

The experimental methods for fabricating heterostructures of complex oxides involve techniques based on pulsed laser deposition, oxide epitaxy (= careful pulsed laser deposition) and molecular beam epitaxy. These methods have made huge progress over the last years and several groups have reported the fabrication of atomically smooth interfaces between various transition metal oxides.

For instance, Ohtomo and coworkers reported on a metallic conductivity in LaTiO$_3$ (LTO)/STO superlattices, a system where both constituents are insulating in bulk [13]. Similarly, metallic transport in LaAlO$_3$ (LAO) /STO [14] heterojunctions was observed and attributed to free electronic carriers at the interface. Thiel *et al.* [15] demonstrated the possibility to tune the carrier density in the latter system by the electric-field-effect in the normal state. Reyren *et al.* [16] demonstrated the superconducting nature of the ground state of the interfacial electron gas and Caviglia *et al.* [17] reported on the use of the electric-field-effect to tune the ground state properties. Other examples involve LaVO$_3$/SrVO$_3$ heterostructures [18], LaVO$_3$/STO interfaces [19] and LAO/LaVO$_3$/LAO quantum wells [20]. Similar to the LTO/STO system, the latter two systems fall into the general class of band-insulator/ Mott-insulator heterostructures. Although the field is still in its infancy, this list is far from being complete and it is expected that many more examples are being studied in the next few years.

## 2.1.2 Classification

Interfaces between two materials can be classified from various point of views [21]. One important characterization is the normal direction of the interface compared to the crystallographic axis. In the oxide examples discussed here, (001) is the most common direction found in the literature and we therefore limit the discussion to this case.[1]

Due to the considerable ionic character of the chemical bonding in transition metal oxide compounds, electrostatic considerations are particularly important for explaining the physical properties of the interface. We distinguish *polar* and *nonpolar* compounds by referring to whether or whether not effectively charged planes alternate in certain high-symmetry directions. For example, A$^{3+}$B$^{3+}$O$_3$ structures are polar in the (001) direction: the (001) planes can be divided into alternating layers of (AO)$^+$ and (BO$_2$)$^-$ just as in III-V semiconductors such as GaAs. On the other hand, the A$^{2+}$B$^{4+}$O$_3$ structure contains formally neutral

---
[1] Hence, we do not touch the interesting dependence of physical properties on the orientation of the interface/surface, see [22] for a theoretical discussion of the superficial metal-to-insulator transition.

## 2.1 Introduction

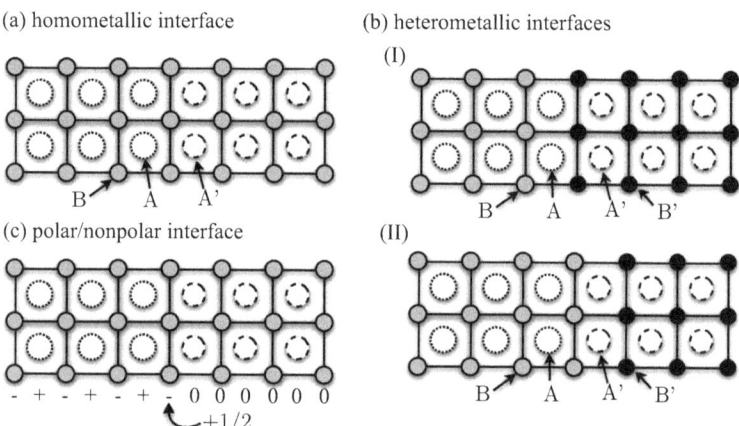

Figure 2.2: Classification of (001) ABO$_3$/A'B'O$_3$ interfaces. (a) Homometallic interface (B = B') and (b) the two possibilities for a heterometallic interface. (c) A polar/nonpolar interface where +1/2 unit charge per unit cell is transfered to the interface layer to reduce the dipole shift.

atomic layers which is analogous to the (001) planes of Si or Ge. Below we give a few examples of polar and nonpolar compounds belonging to the family of perovskites:

- *polar* in (001): LaTiO$_3$, LaAlO$_3$, LaVO$_3$, LaMnO$_3$
- *nonpolar* in (001): SrTiO$_3$, SrMnO$_3$.

In order to minimize the polar discontinuity at the interface between polar and nonpolar materials the system can react by various reconstruction mechanisms, see Fig. 2.2(c). It was argued that in some cases electronic charge transfer is realized which can drive the interfaces between two insulators metallic [23]. This issue will be discussed in more details in Sec. 2.2.

Furthermore, it is useful to distinguish *homometallic* and *heterometallic* structures involving two perovskite materials ABO$_3$ and A'B'O$_3$, see Fig. 2.2. The term *homometallic* refers to heterostructures where A differs from A' but B = B'. Examples of recently studied combinations are STO/ LTO [13] and SrMnO$_3$/ LaMnO$_3$ [24, 25]. Because in these structures the B cation is the same on both sides of the interface, only one type of chemical interface exists. The situation is different in *heterometallic* structures where B differs from B', see Fig. 2.2(b). In case also A differs from A' there exist two types of chemical interfaces, depending on the termination layers at the interface. In polar heterojunctions, they correspond to the n- and p-type interfaces which can have very different physical properties. For example, the AlO$_2$/LaO/TiO$_2$ (n-type) interface between STO and LAO is metallic whereas insulating behavior is observed at the p-type interface AlO$_2$/SrO/TiO$_2$ [23].

## 2.2 Reconstruction and the polar catastrophe

At solid surfaces or solid-solid interfaces the local physical (and chemical) conditions differ considerably from the conditions found in bulk. The system in general responds to these changes by relaxation and reconstruction mechanisms and by introducing defects. In crystals with substantial ionic character, electrostatic boundary conditions are of particular importance in controlling the atomic and electronic structure at interfaces and surfaces. As an illustrative example we first consider the electrostatic properties of polar oxide surfaces [26] and discuss later relaxation and reconstruction mechanisms which are relevant for the oxide heterostructures.

### 2.2.1 Macroscopic electrostatic considerations

Consider a polar compound consisting of a finite number of alternating positively and negatively charged layers, see Fig. 2.3. The relevant macroscopic electrostatic properties can be understood by considering an array of uniformly charged plates separated by $a/2$ where $a$ denotes the lattice spacing between two neighboring B atoms. Without charge reconstruction of the surface layers [case (a)], the charge density is approximated by

$$\rho(z) = \sum_{l=0}^{N_C-1} \sigma \left\{ \delta(z-la) - \delta[z-(l+1/2)a] \right\} \qquad (2.1)$$

## 2.2 Reconstruction and the polar catastrophe

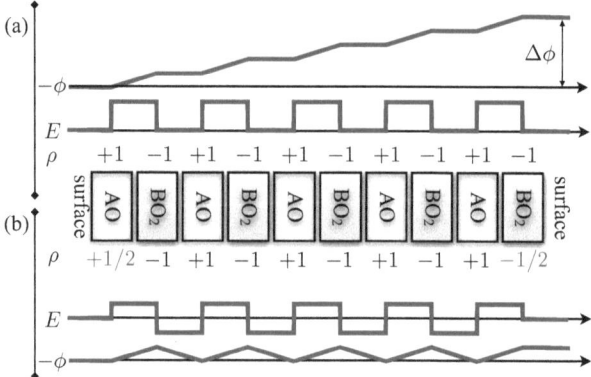

Figure 2.3: Schematic electrostatic diagram for a polar compound showing the charge density $\rho$, the electrical field $E$ and the electrostatic potential $\phi$. (a) Without reconstruction, the difference of the electrostatic potential from one side to the other grows linearly with the number of layers. (b) Reconstructed surfaces avoid the large energy cost of the case (a).

where $\sigma = e/a^2$ with $e > 0$ the unit of charge. $N_C$ denotes the number of negatively (or positively) charged plates. Hence, case (a) corresponds to an array of $N_C$ parallel-plate capacitors with surface charge $\sigma$ and the electric field alternates between 0 and $4\pi\sigma/\epsilon$ where we have introduced an effective dielectric constant $\epsilon$. Inside a single parallel-plate capacitor the electrostatic potential grows linearly whereas it stays constant between neighboring capacitors. As a consequence, an overall dipole shift between the left ($\phi_l$) and the right ($\phi_r$) side of the sample is accumulated:

$$\Delta\phi = |\phi_l - \phi_r| = \frac{4\pi e}{\epsilon a} N_C. \qquad (2.2)$$

Thus, the electrostatic energy in the unreconstructed situation grows linearly with $N_C$. In order to avoid this large energy cost – the so-called *polar catastrophe* – the system reacts by a charge reconstruction of the surfaces, see the case (b) in

Fig. 2.3. By (formally) transferring half of a unit charge per unit cell from the left to the right surface the dipole shift only amounts to

$$\Delta\phi = \frac{2\pi e}{\epsilon a},  \quad (2.3)$$

independent of $N_C$. Of course, similar conclusions are reached when replacing the vacuum on either side by a nonpolar (lattice-matched) material. It follows that for increasing $N_C$ the reconstructed case (b) is energetically always favored. On the other hand, it is possible that a minimal number of unit cells is required to induce the charge compensation at the interface. For instance, Thiel et al. [15] observed that for the n-type interface between STO/LAO a minimal number of 4 unit cells of LAO is required to obtain metallic behavior.

In actual, there are different mechanisms how the charge compensation is realized in a particular system depending, above all, also on the oxygen pressure during the film growth. For the considered transition metal oxide heterostructures there are essentially the following possibilities:

- *Ionic compensation.* For instance,
  - the interdiffusion of the cations (A and A') resulting in a microscopical roughening of the interface or by
  - introducing charged defects, in particular oxygen vacancies.

- *Electronic compensation.* The possibility of realizing mixed valence states in transition metal oxides allows for the transfer of electronic charge to or away from the interface.

While in the ionic compensation mechanism the ideal stoichiometry of the nanostructure is violated, it is essentially preserved when electronic compensation dominates. Noteworthy, because of the lack of mixed-valance states, the possibility of an electronic compensation is not available for conventional polar semiconductor interfaces. The importance of this mechanism was first realized in the context of $K_3C_{60}$ surfaces [27] and soon also for complex oxide heterostructures [23, 28]. For a recent review on conducting interfaces between polar and nonpolar insulating perovskites discussing the different compensation mechanisms in more details we refer to [29].

## 2.2 Reconstruction and the polar catastrophe

### 2.2.2 Relaxation and interface reconstruction

Conceptually independent of the macroscopic electrostatic considerations made above, lattice mismatch, band offsets and broken bonds can lead to a variety of different relaxation and reconstruction mechanisms as well as to the introduction of defects at the interface. A quantitative understanding of all the possible effects is a highly non-trivial task and we refrain from discussing this broad subject in details [21]. Instead, we mention below only a few points which have been put forward in the context of the oxide heterostructures.

**Relaxation**

Even in situations with ideal stoichiometry, lattice relaxation occurs and the lattice constant in the direction perpendicular to the interface varies. In addition, due to the mismatch in the lattice constants of the bulk materials, strain fields are induced. Such effects can influence the dielectric properties near the interface which governs for example the charge leakage of the electrons. Indeed, first principal calculations for the STO/LTO interface show that lattice polarization and relaxation can significantly affect the screening and therefore the details of the charge distribution [30–32].

**Interface reconstruction**

Traditionally, a main issue in surface science is the study of the atomic rearrangement at the surface or interface: the *atomic reconstruction*. For instance, the unit cell of the clean Si(111) surface layer contains 49 atoms and it took decades to show that the $(7 \times 7)$ reconstructed surface is really an equilibrium phase of the clean surface [21]. The ionic compensation including the interdiffusion of the cations at a polar/nonpolar interface discussed above is another example of atomic reconstruction.

In 2004, Okamoto and Millis [28] proposed that *electronic reconstruction* is a further key element for the understanding of the oxide heterostructures: what is the electronic phase at the interface? How does it differ from the bulk phases of either of the two constituents? An important example for the electronic reconstruction mechanism is the metallic behavior of the interface region between insulating materials, possibly triggered by the electronic compensation of the interface dipole as seen in STO/LTO and STO/LAO. But also magnetic correlations [33] and superconductivity [16] were observed which both are absent in the bulk materials.

As a third reconstruction mechanism, the possibility of *orbital reconstruction* was put forward [34]. This mechanism refers to the reconstruction of the orbital occupation and orbital symmetry at the interface layers possibly associated with the formation of covalent bonds.

## 2.3 Theoretical approaches

A variety of theoretical approaches have been considered for the above mentioned systems. In general, we can distinguish between (i) first-principle calculations on the basis of the DFT with suitable simplifications for the exchange-correlation energy, like the LDA [35, 36], and (ii) effective models combined with various many-body techniques, field theoretical approaches and numerical approaches to address electronic correlation effects not included in the LDA [28, 37–45]. Moreover, there is increasing interest in combining first-principle calculations with many-body approaches to obtain realistic electronic structures, most notably the LDA+DMFT approach [46]. A combination of both points of view was also used to study the effect of lattice polarization [30] or the influence of the crystal symmetry on the metallic behavior of STO/LTO superlattices [47]. However, in all approaches a considerable numerical effort is needed to simulate large superlattice units and/or to handle strong local correlations in a spatially non-uniform system. We will comment in more details on the results of these calculations when it is appropriate.

## 2.4 Model of a homometallic heterostructure

Motivated by the experimentally realized STO/LTO heterostructures Okamoto and Millis introduced a generalized Hubbard model to investigate the electronic properties [28, 39–41, 48]. In the following, we concentrate on the simplest version of this model where the relevant orbital degrees of freedom for the Ti-3d electrons are disregarded. Thus, the non-interacting part of the model is sim-

## 2.4 Model of a homometallic heterostructure

ply given by the nearest-neighbor hopping tight-binding Hamiltonian for $s$-like atomic orbitals. As mentioned above, in several publications also more realistic models including the relevant orbital degrees of freedom of the Ti-3$d$ electrons in the tight-binding approximation [28, 40] as well as in combination with band-structure calculations [30, 47] has been considered. A treatment of correlation effects in these multi-orbital models is computationally very expensive and therefore either limited to small system sizes (super unit cell in the direction perpendicular to the interface) or to approximations which are numerically less expensive, like the Hartree(-Fock) approximation, but which can not describe the suppression of the quasiparticle weight due to strong local correlations.

In general, the proximity to a surface or interface can change the electron interaction parameters, the electron bandwidth and the level degeneracy. On the other hand, due to the complexity of the considered system, it is desirable to study idealized models in different parameter regimes and to ask for generic features. We therefore concentrate in this thesis on the electronic properties of a system with perfect lattice match and ideal cubic symmetry. Furthermore, to keep the number of parameters at a minimal level we assume constant values of the interaction and hopping parameters throughout the whole heterostructure.

### 2.4.1 Hubbard heterostructure

We study the geometry of a (001) (homometallic) heterostructure as sketched in Fig. 2.4. We assume a perfect lattice match between the two materials, thereby neglecting aspects related to the lattice relaxation [30]. The microscopic model is given by an extended single-orbital Hubbard model on a cubic lattice

$$H = H_t + H_U + H_{ee} + H_{ei} + H_{ii}. \tag{2.4}$$

Here, the kinetic energy term accounts for nearest-neighbor hopping and the on-site repulsion is modeled by a Hubbard interaction,

$$H_t = -t \sum_{\langle ij \rangle, \sigma} c_{i\sigma}^\dagger c_{j\sigma} + \text{h.c.} \quad \text{and} \quad H_U = U \sum_i n_{i\uparrow} n_{i\downarrow}, \tag{2.5}$$

where $n_{i\sigma} = c_{i\sigma}^\dagger c_{i\sigma}$. The nanoscale structure is defined by the electrostatic potential of $N$ layers of counter ions with a simple positive charge $e$. They sit

*A few basic concepts and a model for correlated heterostructures*

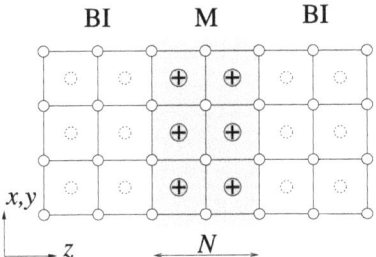

Figure 2.4: Sketch of the heterostructure studied in this thesis. BI is a band insulator and M is a material which can be tuned from a paramagnetic metal to a paramagnetic Mott insulator by varying the onsite repulsion. The heterostructure is defined by $N$ positively charged counter ions sitting in between the electronic sites.

in between the electronic sites of a simple cubic lattice, i.e. on the A sites of the $ABO_3$ structure, and simulate the different electric charge of the $Sr^{2+}$ and $La^{3+}$ ions. Consequently, the conduction electrons are subject to the long-range electron-ion interaction

$$H_{ei} = -E_C \sum_{i,j} \frac{n_i}{|\vec{r}_i - \vec{r}_j^{\,\text{ion}}|} \qquad (2.6)$$

where $\vec{r}_j^{\,\text{ion}}$ denotes the position of the ions, $n_i = n_{i\uparrow} + n_{i\downarrow}$, and we have introduced the parameter $E_C$ controlling the screening length. Furthermore, the long-range electron-electron and ion-ion interaction energies are given by

$$H_{ee} = \frac{E_C}{2} \sum_{i \neq j} \frac{n_i n_j}{|\vec{r}_i - \vec{r}_j|}, \quad H_{ii} = \frac{E_C}{2} \sum_{i \neq j} \frac{1}{|\vec{r}_i^{\,\text{ion}} - \vec{r}_j^{\,\text{ion}}|}, \qquad (2.7)$$

respectively. The number of electrons is fixed by the charge-neutrality condition. Notice that we can formally relate the parameter $E_C$ to an effective dielectric constant $\epsilon = e^2/(E_C a)$ where $a$ is the lattice constant and $e > 0$ the elementary charge. In this interpretation, the factor $1/\epsilon$ takes account of the reduction of the Coulomb force between charges caused by the electronic polarization of the core electron background which we assume to be equal in both compounds.

## 2.4 Model of a homometallic heterostructure

However, for a more realistic description of the screening at the interface a single parameter for the long-range electron-electron interaction is too crude. In fact, the polarization of the lattice dominates the dielectric constant in the considered transition metal oxides and the effect of the relaxation of the lattice near the interface introduces additional parameters in an effective model description [30, 31]. For simplicity, such effects are not considered.

For charge neutral systems, the model (2.4) involves three different parameters, namely $E_C/t$, $N$ and $U/t$. The relative strength of the long-range Coulomb interaction $E_C/t$ mainly affects the self-consistent screening of the ions, as shown below. A realistic value for the experimental system realized by Othomo et al. [13] was estimated to be $E_C = 0.8t$ corresponding to $\epsilon \approx 15$, $a \approx 3.9$ Å and $t \approx 0.3$ eV [28, 39]. The relative onsite repulsion $U/t$ is intrinsically large for Mott insulators. Nevertheless, it is also instructive to consider the weakly interacting limit and we treat this parameter continuously.

### 2.4.2 Thomas-Fermi screening and charge leakage

In order to gain some first insights into the problem of the charge leakage across the interface we shall discuss a Thomas-Fermi approximation in the spirit described in Ref. [37, 38]. For this purpose we temporarily use a continuous description for the electronic charge density $-e\rho(\vec{r})$ and the positive charge density $e\rho_+(\vec{r})$ from the background of the counter ions. In the end we may evaluate $\rho(\vec{r}_i)$ at the position of the electronic lattice sites. The Thomas-Fermi functional for the particle density $\rho(\vec{r})$ is given by

$$F^{\mathrm{TF}}[\rho(\vec{r})] = F_0[\rho(\vec{r})] + \int d^3r V_+(\vec{r})\rho(\vec{r}) + \frac{e^2}{2} \int\int d^3r d^3r' \frac{\rho(\vec{r})\rho(\vec{r}\,')}{|\vec{r}-\vec{r}\,'|} \qquad (2.8)$$

where $F_0(\rho)$ denotes the free-energy of the homogeneous system at particle density $\rho$. The external potential $V_+(\vec{r})$ is generated by the external positive charge density $e\rho_+(\vec{r})$ of the counter-ion layers. In the present case, $F_0(\rho)$ should be calculated from the homogeneous three-dimensional Hubbard model. Variation of this functional under the constraint of charge neutrality $\int \rho(\vec{r})d^3r = \int \rho_+(\vec{r})d^3r$ leads to the basic Thomas-Fermi equation

$$\mu_0[\rho(\vec{r})] - e\phi(\vec{r}) = \mu = \mathrm{const.} \qquad (2.9)$$

Here, the electrostatic potential $\phi(\vec{r})$ is determined by Poisson's equation

$$\Delta\phi(\vec{r}) = -\frac{4\pi e}{\epsilon}[\rho_+(\vec{r}) - \rho(\vec{r})] \qquad (2.10)$$

and

$$\mu_0(\rho) = \frac{1}{V}\frac{\partial F_0(\rho)}{\partial \rho} \qquad (2.11)$$

denotes the chemical potential of the homogeneous system of volume $V$.

For the present discussion it is sufficient to assume that the charge density depends only on the $z$-coordinate and is uniform within a single plane. The Thomas-Fermi problem then reduces to

$$\frac{d^2\phi(z)}{dz^2} = \frac{4\pi e}{\epsilon}[\rho(z) - \rho_+(z)], \quad \mu_0[\rho(z)] - e\phi(z) = \mu. \qquad (2.12)$$

Taking two times the derivative of the second equation of (2.12) with respect to $z$ yields a single differential equation

$$\frac{d\mu_0}{d\rho}\frac{d^2\rho}{dz^2} + \frac{d^2\mu_0}{d\rho^2}\left(\frac{d\rho}{dz}\right)^2 = \frac{4\pi e^2}{\epsilon}[\rho(z) - \rho_+(z)]. \qquad (2.13)$$

In addition, in order to determine and eventually solve (2.13), the chemical potential $\mu_0(\rho)$ as function of $\rho$ is required. For the 3d Hubbard model $\mu_0(\rho)$ is unknown and one has to rely on approximate expressions. In the following we will comment on some simple approximations.

**Toy model**

Let us first discuss the situation where the density is varying slowly on the typical length scale $1/k_F$ of electrons at the Fermi energy. We can then assume that

$$\frac{d\mu_0}{d\rho} = \text{const.} \qquad (2.14)$$

in the considered density range. This yields the well-known Thomas-Fermi screening length[2]

$$\lambda_{\text{TF}} = \sqrt{\frac{\epsilon}{4\pi e^2}\frac{d\mu_0}{d\rho}}. \qquad (2.15)$$

---

[2]For the considered interface between a Mott and a band-insulator a natural extension would be to introduce two length scales associated with the two different materials.

## 2.4 Model of a homometallic heterostructure

In order to illustrate the dependence of $\lambda_{\rm TF}$ on $U$ we use a simplified relation between the density $\rho$ and the chemical potential $\mu_0$, namely, a piecewise linear relation:

$$\mu_0(\rho) = \frac{1}{2} \times \begin{cases} -W + \rho a^3[W + U - \Delta(U)], & \rho a^3 < 1; \\ U, & \rho a^3 = 1; \\ [\Delta(U) + U] + (\rho a^3 - 1)[W + U - \Delta(U)], & \rho a^3 > 1. \end{cases} \quad (2.16)$$

Here, $\Delta(U)$ is the Mott-Hubbard gap and $W$ denotes the bandwidth. With this approximation we find

$$\frac{d\mu_0}{d\rho} = \frac{W + U - \Delta(U)}{2} a^3$$

for $\rho a^3 \neq 1$ and consequently

$$\lambda_{\rm TF} = \sqrt{\frac{W + U - \Delta(U)}{8\pi E_C}} a \quad (2.17)$$

and $E_C = e^2/(a\epsilon)$ as above. In the weakly correlated regime, $U \ll W$, $\Delta = 0$ and $\lambda_{\rm TF}$ increases by increasing $U$. On the other hand, for $U \gg W$ the gap is $\Delta \approx U$ and $\lambda_{\rm TF}$ is independent of $U$ and given by

$$\lambda_{\rm TF} \approx \sqrt{\frac{W}{8\pi E_C}} a \sim \sqrt{\frac{t}{E_C}} a. \quad (2.18)$$

Moreover, $\lambda_{\rm TF}$ is of the order of the lattice constant $a$. We note here that the dependence of the density profile on $U$ obtained by a more complete treatment of the screening as described in Chap. 4 is consistent with the conclusions reached from the toy model (2.16).

The assumption of a constant screening length yields for a quantum well geometry with a well of width $d_{\rm MI} = Na$ the following density profile

$$n(z) \equiv \rho(z)a^3 = \begin{cases} \sinh\left(\frac{d_{\rm MI}}{2\lambda_{\rm TF}}\right) e^{z/\lambda_{\rm TF}}, & z < -d_{\rm MI}/2; \\ 1 - e^{-d_{\rm MI}/(2\lambda_{\rm TF})} \cosh(z/\lambda_{\rm TF}), & |z| \leq d_{\rm MI}/2; \\ \sinh\left(\frac{d_{\rm MI}}{2\lambda_{\rm TF}}\right) e^{-z/\lambda_{\rm TF}}, & z > d_{\rm MI}/2. \end{cases} \quad (2.19)$$

## A few basic concepts and a model for correlated heterostructures

In particular, the density $n_{l=0} \equiv n(z=0)$ in the center of the heterostructure[3] depends exponentially on the width of the quantum well

$$n_0 = 1 - e^{-d_{\mathrm{MI}}/(2\lambda_{\mathrm{TF}})}.$$

Using the above expression we can estimate the hopping renormalization factor in the center of the heterostructure form the Gutzwiller approximation (GA) [100] for $U \gg W$ and $n \approx 1$:[4]

$$z_0^2 \approx 2[1-n_0] = 2\exp\left(-\frac{d_{\mathrm{MI}}}{2\lambda_{\mathrm{TF}}}\right) = 2\exp\left(-cN\sqrt{\frac{E_C}{t}}\right).$$

Here, $c$ is a constant of order unity. In the local Fermi liquid interpretation of the GA result, $z_0^2$ corresponds to the local quasiparticle weight in the center of the quantum well. In the strongly interacting regime it is thus exponentially small in $N$. This is again consistent with the results of Chap. 4 where similar aspects are discussed in more details.

**Atomic limit**

Let us now look at the limit of vanishing hopping amplitude $t$, the *atomic limit* at finite temperature $T$. In this limit, the relation between chemical potential and particle density is given by [49] ($\beta = 1/k_B T$)

$$n = \frac{2\xi + 2\xi^2 e^{-\beta U}}{1 + 2\xi + \xi^2 e^{-\beta U}}, \quad \xi = e^{\beta \mu_0}, \tag{2.20}$$

which yields

$$\mu_0(n) = \frac{1}{\beta}\ln\xi = k_B T \ln\left[\frac{n-1+\sqrt{(1-n)^2 + n(2-n)e^{-\beta U}}}{e^{-\beta U}(2-n)}\right]. \tag{2.21}$$

In the *strongly correlate regime*, $\beta U \gg 1$, and for $n \leq 1$ we find

$$\mu_0(n) = -k_B T \ln\left[\frac{2(1-n)}{n}\right] \tag{2.22}$$

---

[3]Here, we assume $N$ even such that there is a single layer at $z=0$ in the center of the quantum well. For $N$ odd there are two central layers at $z = \pm a/2$ but similar expressions are found.

[4]See also Eq. (3.21) and the discussion of Sec. 3.2.

## 2.4 Model of a homometallic heterostructure

and hence
$$\beta\frac{d\mu_0}{dn} = \frac{1}{n(1-n)}, \quad \beta\frac{d^2\mu_0}{dn^2} = \frac{2n-1}{n^2(1-n)^2}. \tag{2.23}$$

Introducing the dimensionless quantities
$$n(\zeta) = a^3\rho(\zeta\lambda_{\text{TF}}), \quad \zeta = \frac{z}{\lambda_{\text{TF}}} \tag{2.24}$$

where $a$ is the lattice parameter and
$$\lambda_{\text{TF}} = \sqrt{\frac{k_B T \epsilon a}{4\pi e^2}} a = \sqrt{\frac{k_B T}{4\pi E_C}} a \tag{2.25}$$

the *characteristic length-scale of the Thomas-Fermi problem in the atomic limit* we eventually obtain the following non-linear differential equation
$$n(1-n)n'' + (2n-1)(n')^2 = n^2(1-n)^2(n-n_+). \tag{2.26}$$

To leading order, we find that in the band-insulator the charge density vanishes as $\rho(z) \sim e^{-z/\lambda_{\text{TF}}}$. Note that (2.26) has the same form in the band and in the Mott insulator. Indeed, the electron density approaches half-filling in the Mott insulator over the same length-scale $\lambda_{\text{TF}}$.

It is interesting to compare this result with the *weakly interacting regime*. Performing a similar calculation we find
$$2n(2-n)n'' + 4(n-1)(n')^2 = n^2(2-n)^2(n-n_+). \tag{2.27}$$

In the band-insulator the characteristic length-scale is reduced to $\lambda_{\text{TF}}/\sqrt{2}$.

For numerical computations it is convenient to use again the discretized geometry of equally spaced layers instead of the continuum description. The solution of the Thomas-Fermi equation (2.9) is then found by solving a set of nonlinear equations[5]
$$n_l = \frac{2\xi}{2\xi + \exp(-e\beta\phi_l)} \tag{2.28}$$

where
$$-e\phi_l = V_l + \sum_{l'} W_{ll'} n_{l'} \tag{2.29}$$

---
[5] Here we adapt the limit $U \to \infty$ for simplicity.

is obtained from the solution of Poisson's equation (2.10), see also Sec. 4.3.3. In Eq. (2.28), $\zeta$ is determined by the charge neutrality condition for a given potential $\phi_l$:

$$N = \sum_l \frac{2\xi}{2\xi + \exp(-e\beta\phi_l)}. \qquad (2.30)$$

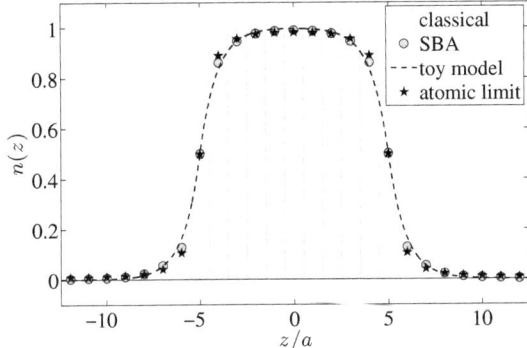

Figure 2.5: Comparison of the density distribution obtained within different approximations for a quantum well of width $N = 10$. The parameters in the SBA solution are $U_r = 22t$ and $E_C = 0.8t$. For the toy-model solution we have $\lambda_{\text{TF}} = 0.9a$ and for the solution in the atomic limit $E_C = 0.8k_BT$ and $U \to \infty$. Also shown is the "classical" charge distribution obtained in the parallel plate capacitor model of Sec. 2.2.1.

**Comparison**

In Fig. 2.5 we compare the density profile of a BI/MI/BI quantum well with $N = 10$ counter-ion layers obtained from the different approaches discussed in this chapter. The electronic density distribution obtained from the "classical" parallel-plate capacitor model of Sec. 2.2.1 is shown by the shaded area. Note that because the location of the counter-ion layers are shifted by half of a unit

## 2.4 Model of a homometallic heterostructure

cell with respect to the electronic layers there are approximately quarter filled interface layers in order to reduce the dipole energy. This is similar to the situation discussed in Sec. 2.2.1. The inclusion of screening leads to charge leakage across the interface which predominantly affects the layers next to the interface layer. We show the results obtained from the toy-model expression (2.19) with $\lambda_{TF} = 0.9a$, from the atomic limit with $E_C = 0.8k_B T$ and $U \to \infty$ as well as the density profile obtained within the full Kotliar and Ruckenstein slave-boson mean-field approximation (SBA) treatment discussed in Chap. 4. In conclusion, the macroscopic electrostatic constraints are important for the stabilization of the interface layers. Furthermore, screening affects the charge leakage across the interface and the characteristic screening length is of the order of the lattice spacing.

# Chapter 3

# Slave-particle methods for strong correlation physics

> *Slave-particle methods are semi-analytical methods widely used to handle strong correlation physics. In this chapter we give a brief overview over commonly used slave-particle representations for fermion lattice models with focus on the four slave-boson formulation of Kotliar and Ruckenstein. At half-filling, a substantial simplification is obtained by introducing pseudospin variables which allows us to discuss the full one-particle spectral density and characteristics of the Mott transition in three dimensions. Eventually, we elucidate the close physical and formal connection to dynamical mean-field theory.*

## 3.1 A brief overview

Using auxiliary fields to describe electronic degrees of freedom has a long history and, depending on the given problem, many different variants can be found in the literature. The most popular used in fermion lattice models are briefly discussed below. In this section, we partly follow Frésard and Wölfle [50]. To illustrate the general principle let us consider the *local* Fock space for the Fermi creation and annihilation operators $c_\sigma^{(\dagger)}$, $\sigma = \uparrow, \downarrow$, defined by the four occupation number

## 3.1 A brief overview

states
$$|e\rangle \equiv |0\rangle, \quad |\uparrow\rangle \equiv c_\uparrow^\dagger |0\rangle, \quad |\downarrow\rangle \equiv c_\downarrow^\dagger |0\rangle, \quad |d\rangle \equiv c_\uparrow^\dagger c_\downarrow^\dagger |0\rangle. \quad (3.1)$$

A substantial simplification is obtained in the limit $U \to \infty$ which projects out the doubly occupied state $|d\rangle$. However, for the subsequent discussion we keep the value of $U$ finite.

### 3.1.1 Schwinger-Wigner representation

The slave-particle representations were pioneered in the context of spin models by Holstein and Primakoff [51] and by Schwinger and Wigner [52]. In the Schwinger-Wigner representation, a spin-$S$ operator is written in terms of two Bose creation and annihilation operators $b_\alpha^{(\dagger)}$:

$$S^a = \frac{1}{2} \sum_{\alpha\beta} b_\alpha^\dagger \sigma_{\alpha\beta}^a b_\beta, \quad \sum_\alpha b_\alpha^\dagger b_\alpha = 2S. \quad (3.2)$$

Here, $\sigma^a$ are the Pauli matrices. This representation was used by Jayaprakash *et al.* [53] and Yoshioka as a starting point for fermion lattice models: introducing additional spinless fermions $e^{(\dagger)}$ and $d^{(\dagger)}$ for empty and doubly occupied sites, (3.2) is generalized to express the Hubbard operators $X_{ab} = |a\rangle\langle b|$ in terms of the slave-particles [54]. The local states (3.1) are represented by

$$|\sigma\rangle = b_\sigma^\dagger |\Omega\rangle, \quad |e\rangle = e^\dagger |\Omega\rangle, \quad |d\rangle = d^\dagger |\Omega\rangle, \quad (3.3)$$

with a fake vacuum vector $|\Omega\rangle$. A constraint is introduced to eliminate unphysical states

$$e^\dagger e + \sum_\sigma b_\sigma^\dagger b_\sigma + d^\dagger d = 1. \quad (3.4)$$

Alluding to the auxiliary $e$ and $d$ fermions, the representation defined by (3.3) and (3.4) is often called the *slave-fermion* representation. It has been used to study the phase diagram for the resonating valence bond state and the spin order in the context of the high-$T_c$ superconductors [53, 55].

### 3.1.2 Barnes-Coleman representation

A different representation of the local states was introduced by Barnes [56] and later by Coleman [57, 58] and Read and Newns [59, 60] to study the Anderson and the Coqblin-Schrieffer model of a magnetic impurity in a metal. In this approach one uses Bose operators $e^{(\dagger)}$ and $d^{(\dagger)}$ for the empty (holon) and the doubly occupied (doublon) states and Fermi operators $f_\sigma^{(\dagger)}$ for the singly occupied (spinon) states

$$|e\rangle = e^\dagger|\Omega\rangle, \quad |\sigma\rangle = f_\sigma^\dagger|\Omega\rangle, \quad |d\rangle = d^\dagger|\Omega\rangle, \quad (3.5)$$

$$e^\dagger e + \sum_\sigma f_\sigma^\dagger f_\sigma + d^\dagger d = 1. \quad (3.6)$$

Due to the fact that the Bose degrees of freedom are "slaved" to the Fermi degrees of freedom, this representation is known under the name of *slave-boson representation*. Beside the implementation for the single-impurity Anderson model it was also widely used to study the periodic Anderson model and the *t-J* model [61] in the context of the high-$T_c$ materials. In particular, the extension to an SU(2) invariant formulation by Wen and Lee [62, 63] has become an important framework for theoretical investigations related to the underdoped cuprates.

### 3.1.3 Kotliar-Ruckenstein representation

The above mentioned representations use either Bose operators for the spin and Fermi operators for the charge degrees of freedom, or vice versa. Kotliar and Ruckenstein [64] introduced yet another representation of the local states which has the advantage of treating spin and charge degrees of freedom on an equal footing. When applied to the single-band Hubbard model, the simplest mean-field treatment (SBA) reproduces the results of the Gutzwiller approximation (GA) [65], particularly, it describes a metal-insulator transition at half filling [66], see Sec. 3.2. This representation is sometimes called the *four slave-boson representation* in order to distinguish it from the Barnes-Coleman formulation. Indeed, four Bose creation and annihilation operators $e^{(\dagger)}$, $p_\sigma^{(\dagger)}$ and $d^{(\dagger)}$ as well as two Fermi operators $f_\sigma^{(\dagger)}$ are introduced such that the local occupation number states (3.1) are written as

$$|e\rangle = e^\dagger|\Omega\rangle, \quad |\sigma\rangle = p_\sigma^\dagger f_\uparrow^\dagger|\Omega\rangle, \quad |d\rangle = d^\dagger f_\uparrow^\dagger f_\downarrow^\dagger|\Omega\rangle. \quad (3.7)$$

The constraints to project out unphysical states are given by

$$e^\dagger e + \sum_\sigma p_\sigma^\dagger p_\sigma + d^\dagger d = 1, \quad d^\dagger d + p_\sigma^\dagger p_\sigma = f_\sigma^\dagger f_\sigma. \quad (3.8)$$

There are generalizations of this representation to explicitly preserve the spin-rotation invariance [67] and/or the charge-rotation invariance [50]. Similarly, for multi-orbital systems, orbital-rotation invariance can be incorporated [68]. We used the mean-field approximation of this approach to study physical properties of correlated heterostructures [69–71]. This is the subject of Chapter 4 and 5 of this thesis. To provide a basis, we consider the mean-field approximation in more details in the following section.

## 3.2 Four boson mean-field approximation

The self-consistency equations of the Kotliar and Ruckenstein slave-boson mean-field approximation (SBA) are most conveniently derived in the path integral representation. As an illustration of the formalism we discuss the single orbital Hubbard model in Sec. 3.2.1. Furthermore, a few results concerning the Mott transition in the two-band Hubbard model with bands of different widths are presented in Sec. 3.2.2. These results show the ability of the method to simultaneously access metallic and insulating properties in the orbital-selective Mott insulator.

### 3.2.1 Formalism for the single orbital Hubbard model

We follow the original work of Kotliar and Ruckenstein and consider the single-orbital Hubbard Hamiltonian

$$H_{\text{Hub}} = -t \sum_{\langle i,j \rangle,\sigma} \left( c_{i\sigma}^\dagger c_{j\sigma} + \text{h.c.} \right) + U \sum_i n_{i\uparrow} n_{i\downarrow}. \qquad (3.9)$$

Here, the hopping is assumed to take place between neighboring sites $i$ and $j$. However, as we will see below, on the mean-field level only the non-interacting density of states enters.

### Slave-boson Hamilton operator

Using the slave-boson representation (3.7) we write the single orbital Hubbard Hamiltonian as

$$H^{\text{sb}} = -t \sum_{\langle i,j \rangle, \sigma} \left( z_{i\sigma}^\dagger z_{j\sigma} f_{i\sigma}^\dagger f_{j\sigma} + \text{h.c.} \right) + U \sum_i d_i^\dagger d_i. \quad (3.10)$$

Due to the redundancy of the enlarged Fock space the choice of the $z$-operators in (3.10) is to some extend arbitrary. We may use the form

$$z_{i\sigma} = e_{i\sigma}^\dagger R_{i\sigma} p_{i\sigma} + p_{i\bar\sigma}^\dagger R_{i\sigma} d_i, \quad z_{i\sigma}^\dagger = e_{i\sigma} R_{i\sigma} p_{i\sigma}^\dagger + p_{i\bar\sigma} R_{i\sigma} d_i^\dagger. \quad (3.11)$$

In the simplest case, $R_{i\sigma} = 1$. However, in order to reproduce the correct $U = 0$ limit on the mean-field level, Kotliar and Ruckenstein [64] suggested to use

$$R_{i\sigma}^{\text{KR}} =: \frac{1}{\sqrt{1 - d_i^\dagger d_i - p_{i\sigma}^\dagger p_{i\sigma}}} \frac{1}{\sqrt{1 - e_i^\dagger e_i - p_{i\bar\sigma}^\dagger p_{i\bar\sigma}}} : \quad (3.12)$$

where $:(\ldots):$ means normal ordering. Alternatively, one can choose a suitable "linearization" [72]

$$R_{i\sigma}^{\text{AS}} = [1 + x(d_i^\dagger d_i + p_{i\sigma}^\dagger p_{i\sigma})][1 + x(e_i^\dagger e_i + p_{i\bar\sigma}^\dagger p_{i\bar\sigma})]. \quad (3.13)$$

where $x = 2(\sqrt{2} - 1)$ is such that the $U = 0$ limit is recovered at the mean-field level. Note that the Gaussian fluctuations around the saddle-point crucially depend on the choice of the $z$-operators, see e.g. [72–75]. However, on the mean-field level this subtle issue is irrelevant and we adopt the original form (3.12).[1] In any case, in the physical subspace defined by the constraints (3.8), the slave-boson Hamiltonian (3.10) is equivalent to the original Hamiltonian (3.9).

### Slave-boson theory on the functional integral level

For the functional integral representation we start by writing the grand canonical partition function as

$$\mathcal{Z}^{\text{sb}}(\beta, \mu) = \text{Tr}\left[e^{-\beta(H^{\text{sb}} - \mu N)}\mathcal{P}\right] \quad (3.14)$$

---

[1] Some aspects of fluctuations are discussed in Sec. 3.3 using a simplified pseudospin formulation.

## 3.2 Four boson mean-field approximation

where the trace extends over the full enlarged Hilbert space and $\mathcal{P}$ projects onto the physical subspace. Explicitly, we may express the projector as

$$\mathcal{P} = \mathcal{P}_I \mathcal{P}_\uparrow \mathcal{P}_\downarrow, \quad \mathcal{P}_\sigma = \prod_i \mathcal{P}_\sigma^{(i)}, \quad \mathcal{P}_I = \prod_i \mathcal{P}_I^{(i)} \quad (3.15)$$

where ($\beta = 1/T$)

$$\begin{aligned}
\mathcal{P}_I^{(i)} &= \int_{-\pi T}^{\pi T} \frac{d\eta_i}{2\pi T} \exp\left[-i\beta\eta_i \left(e_i^\dagger e_i + \sum_\sigma p_{i\sigma}^\dagger p_{i\sigma} + d_i^\dagger d_i - \mathbf{1}_i\right)\right], \\
\mathcal{P}_\sigma^{(i)} &= \int_{\pi T}^{\pi T} \frac{d\lambda_{i\sigma}}{2\pi T} \exp\left[-i\beta\lambda_{i\sigma} \left(p_{i\sigma}^\dagger p_{i\sigma} + d_i^\dagger d_i - f_{i\sigma}^\dagger f_{i\sigma}\right)\right].
\end{aligned}$$

The above projectors enforce the constraints (3.8) on each lattice site $i$. Since the physical subspace is conserved under the evolution governed by (3.10) we can combine the exponentials in (3.14). We then use a functional integral representation for the trace to find

$$\mathcal{Z}^{\text{sb}} = \int D(d_i^*, d_i) D(p_{i\sigma}^*, p_{i\sigma}) D(e_i^*, e_i) \int_{-\pi T}^{\pi T} \prod_i \frac{d\eta_i}{2\pi T} \prod_{i\sigma} \frac{d\lambda_{i\sigma}}{2\pi T} e^{-S[d,p,e,\eta,\lambda]} \quad (3.16)$$

where

$$S[d,p,e,\lambda,\eta] = \int_0^\beta d\tau (L_B + L_F) \quad (3.17)$$

with

$$\begin{aligned}
L_B &= \sum_i \Big[d_i^*(\tau) \left(\partial_\tau + i\eta_i - i\lambda_{i\uparrow} - i\lambda_{i\downarrow} + U\right) d_i(\tau) \\
&\quad + \sum_\sigma p_{i\sigma}^*(\tau) \left(\partial_\tau + i\eta_i - i\lambda_{i\sigma}\right) p_{i\sigma}(\tau), \\
&\quad + e_i^*(\tau)(\partial_\tau + i\eta_i) e_i(\tau) - i\eta_i\Big] \quad (3.18)
\end{aligned}$$

$$L_F = \sum_{ij\sigma} f_{i\sigma}^+(\tau) \left[\delta_{ij}(\partial_\tau - \mu + i\lambda_{i\sigma}) + t_{ij} z_{i\sigma}^* z_{j\sigma}\right] f_{i\sigma}(\tau). \quad (3.19)$$

**Saddle-point approximation**

The slave-boson mean-field approximation is obtained from (3.16) by a saddle-point approximation of the functional integral over the slave bosons and the Lagrange multipliers. It is possible to discuss unbiased saddle-point solutions when considering finite clusters [76]. For bulk systems one has to consider restricted saddle points: however, a variety of different phases can be discussed including magnetic and spiral [77, 78] as well as stripe phases [79].[2] The simplest assumption of a uniform paramagnetic phase yields at $T = 0$ and for a fixed particle density $n$ the following energy per site:

$$E_G(d, \lambda) = 2 \int d\varepsilon\, \rho_\sigma(\varepsilon)(z^2\varepsilon + \lambda)f_0(z^2\varepsilon + \lambda) + Ud^2 - \lambda n \quad (3.20)$$

where $\rho_\sigma(\varepsilon)$ is the non-interacting density of states per spin projection, $f_T(E) = (1 + e^{E/T})^{-1}$ is the Fermi-Dirac distribution and the hopping renormalization factor is given by

$$z(n, d) = \frac{\sqrt{(1 - n + d^2)(n - 2d^2)} + d\sqrt{n - 2d^2}}{\sqrt{n(1 - n/2)}}. \quad (3.21)$$

Maximizing $E_G(d, \lambda)$ with respect to $\lambda$ results in an energy functional $E_{GA}(d)$ of the double occupancy alone. In fact, $E_{GA}(d)$ is the energy functional known from the GA, see below. At half filling $n = 1$, the minimization of $E_{GA}$ with respect to $d$ yields

$$d^2 = \begin{cases} \frac{1}{4}(1 - u) & \text{for } u \leq 1, \\ 0 & \text{for } u > 1; \end{cases} \quad \text{and} \quad z^2 = \begin{cases} 1 - u^2 & \text{for } u \leq 1, \\ 0 & \text{for } u > 1. \end{cases} \quad (3.22)$$

where $u = U/U_c$ with

$$U_c = 16 \left| \int_{-\infty}^{0} d\varepsilon\, \varepsilon \rho_\sigma(\varepsilon) \right|. \quad (3.23)$$

The renormalization factor $z^2$ is obviously related to an enhancement of the effective mass: $m^*/m = 1/z^2$. The divergence of $m^*$ at the critical interaction strength $U_c$ signals the transition to a localized state. This is the Brinkman-Rice

---
[2]Note that depending on the considered phase it is crucial to rely on a spin-rotation-invariant formulation [67].

## 3.2 Four boson mean-field approximation

[66] picture of the Mott transition from a paramagnetic metal to a paramagnetic insulator. On the other hand, $z^2$ can be identified with the quasiparticle weight $Z$, its vanishing indicates the loss of low-lying quasiparticle excitations. This one-to-one correspondence between effective mass and quasiparticle weight is a general feature of a momentum independent self-energy $\Sigma(\omega)$ and is therefore rigorously recovered in the infinite dimensional system [80]. In finite dimensions the divergence of the effective mass is usually cut-off but the relation $m^*/m \sim 1/Z$ is roughly fulfilled for many three dimensional compounds [11], see also Sec. 3.3.5. Nevertheless, we mention that a different behavior is observed in the layered high-$T_c$ cuprates when approaching the Mott-insulating parent compound by reducing the doping: $Z$ vanishes but the Fermi velocity in the nodal direction stays approximatively constant [81, 82].

Obviously, the saddle-point approximation always reduces the complicated many-body problem to a system of non-interacting fermionic quasiparticles with self-consistently renormalized properties. Thus, the saddle-point solution is a straight-forward realization of a Fermi liquid. The above described metal-insulator transition is a particular instability of this state. Of course, depending on the details of the non-interacting dispersion and the lattice topology, other instabilities, like a spin-density wave, are favored. Indeed, it is possible to calculate various Landau parameters starting from the microscopic model and to study the associated instabilities, see e.g. [83, 84].

**Relation to the Gutzwiller variational principle**

As mentioned above, the SBA reproduces in many situation the results of the Gutzwiller approximation (GA). The GA is actually a subsequent approximation performed in another powerful approach to Hubbard-type lattice models: the variational principle using so-called *Gutzwiller projected wave functions* [65]. In this scheme, an approximation $|\Psi\rangle$ for the ground state wave function is constructed by applying the Gutzwiller projector $\mathcal{P}_G$ to a Slater-determinant $|\Psi_0\rangle$,

$$|\Psi\rangle = \mathcal{P}_G |\Psi_0\rangle. \tag{3.24}$$

In general, both the projector as well as $|\Psi_0\rangle$ contain variational parameters which are determined by minimizing $\langle\Psi|H|\Psi\rangle/\langle\Psi|\Psi\rangle$ using variational Monte Carlo methods. In the simplest situation, $\mathcal{P}_G$ projects out configurations in $|\Psi_0\rangle$

which contain doubly occupied sites (with a probability depending on $U$). In its original work, Gutzwiller considered a projected Fermi sea to study magnetic metals. With Anderson's observation that a resonating valence bond (RVB) state could be formally generated by projecting a Bardeen-Cooper-Schrieffer (BCS) pair superconducting state [85], the use of Gutzwiller projected wave functions has become an important tool in the study of the high-$T_c$ cuprates, see [82] for a recent review.

The Gutzwiller approximation (GA) consists of an analytical approximation to evaluate expectation values of arbitrary operators in the projected state. It can be controlled by a $1/d$ expansion [86]. Explicitly, we approximate the expectation value of an operator $\mathcal{O}$ by

$$\frac{\langle \Psi | \mathcal{O} | \Psi \rangle}{\langle \Psi | \Psi \rangle} \approx g_{\mathcal{O}} \langle \Psi_0 | \mathcal{O} | \Psi_0 \rangle \qquad (3.25)$$

where the renormalization factor $g_{\mathcal{O}}$ is given by the probability for the physical process under consideration to occur in the projected wave function divided by the probability for such a process to occur in the preprojected wave function. Although the starting point for the GA is very different from the slave-boson formulation, it is important to note that the energy functional obtained in the SBA is in most situations identical with the one obtained in the GA for a projected Fermi sea.[3] On the other hand, the Gutzwiller approximation for projected BCS states has led to the renormalized mean-field theory (RMFT) for the underdoped cuprates [87] which is not directly related to the SBA discussed in the previous paragraph.

### 3.2.2 The Mott transition in the two-band Hubbard model

We illustrate the physical content of the saddle-point approximation for the paramagnetic metal-insulator transition in the two-band Hubbard model with bands of different widths [88]. For an Ising-like Hund's coupling the slave-boson approach of Kotliar and Ruckenstein is easily generalized to the multi-band system. Additional spin-flip and pair-hopping terms considerably complicate the problem and an extension which is rotational invariant in the spin and orbital degrees of

---

[3]Nevertheless, this equivalence can break down in inhomogeneous systems, see Sec. 4.6.2.

## 3.2 Four boson mean-field approximation

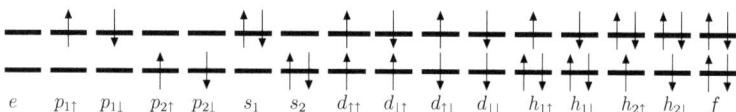

$e \quad p_{1\uparrow} \quad p_{1\downarrow} \quad p_{2\uparrow} \quad p_{2\downarrow} \quad s_1 \quad s_2 \quad d_{\uparrow\uparrow} \quad d_{\uparrow\downarrow} \quad d_{\downarrow\uparrow} \quad d_{\downarrow\downarrow} \quad h_{1\uparrow} \quad h_{1\downarrow} \quad h_{2\uparrow} \quad h_{2\downarrow} \quad f$

Figure 3.1: The slave-boson fields introduced to represent the states of the local occupation number basis in the two-band Hubbard model.

freedom has to be adopted [68]. We focus here on the Ising-like Hund's coupling for which it is sufficient to introduce a slave-boson operator associated with each of the 16 possible local occupation states, see Fig. 3.1.

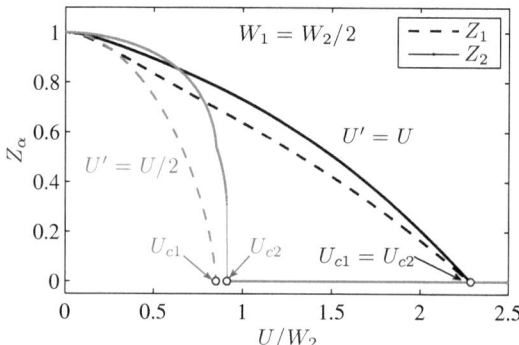

Figure 3.2: The quasiparticle weight in the two-band Hubbard model as obtained by the SBA in the paramagnetic phase. The band width of the narrow band is half of the width of the wide band, $W_1 = W_2/2$. Furthermore, we adopt the relation $U = U' + 2J$ between the onsite interaction parameters.

For simplicity, we have used box density of states to characterize the non-interacting bands of width $W_1$ and $W_2 = 2W_1$. In Fig. (3.2) we show the quasiparticle weight $Z_\alpha$ associated with the two bands $\alpha = 1, 2$ as a function of the onsite (intraorbital) repulsion $U$. We have used the relation $U = U' + 2J$ between the intraorbital interaction $U$, the interorbital repulsion $U'$ and the (Ising)

Hund's coupling $J$. Figure (3.2) shows that for $U = U'$ there is a joint transition at $U_{c1} = U_{c2}$ involving both bands. In this case, there is an enlarged local symmetry with six degenerate two-electron onsite congurations: four spin congurations with one electron in each orbital and two spin singlets with both electrons in one of the two orbitals [89]. As a consequence, charge fluctuations are more pronounced and the metallic phase is more stable compared to cases where $U' \neq U$.

For $U' = U/2$, the narrow band undergoes an orbital-selective Mott transition (OSMT) at $U_{c1}$ and the quasiparticle weight of the wider band remains finite up to $U_{c2}$ where it jumps to zero. Within the SBA we find a first order transition in the wider band. The values of $U_{c1}$ and $U_{c2}$ are relatively close together which is a consequence of the Ising-type Hund's coupling. If an isotropic Hund's coupling is considered, the orbital-selective Mott-phase is stabilized [90]. The nature of the paramagnetic Mott transition, in particular the possibility of an OSMT and the role of the Hund's coupling, has also been extensively studied by use of the dynamical mean-field theory [89, 91–93]. This interest in the OSMT was triggered by the proposal that parts of the intriguing phase diagram of the $Ca_{2-x}Sr_xRuO_4$ alloy could be explained under the assumption of an OSMT involving two of the three partially filled $d$-electron bands [94–96].

## 3.3 A slave-spin theory for the particle-hole symmetric Hubbard model

The mean-field approximation outlined in Sec. 3.2 is incomplete in that it focuses only on the coherent part of the wave function (i.e. the part which is described by renormalized quasiparticles). In other words, the local moments which are formed as a consequence of strong local correlations as well as the atom-like features describing transitions between different local charging states are missing. In this section we discuss a minimal formulation of the paramagnetic phase at half-filling which reproduces the SBA (or GA) result in the simplest mean-field treatment but allows us to go further by treating the low and high energy part on an equal footing [97]. This formulation offers a convenient way to obtain the lower and upper Hubbard bands and to discuss characteristics of the Mott

## 3.3 A slave-spin theory for the particle-hole symmetric Hubbard model

transition in three dimensions. At least at half filling, the use of new pseudospin variables removes some of the difficulties encountered in a systematic expansion around the saddle-point of the Kotliar-Ruckenstein action (3.17) [73, 74, 98, 99].

We first introduce the general formulation of the problem using pseudospin variables in Sec. 3.3.1. We then discuss the mean-field approximation which consists of decoupling pseudospin and fermion degrees of freedom in Sec. 3.3.2. Furthermore, we investigate the consequences of two subsequent approximations made for the pseudospin problem: (i) the single-site molecular field approximation and (ii) the use of Schwinger bosons to treat the fluctuations around the classical ground state.

### 3.3.1 Model and method

We shall consider the single band Hubbard model at half filling ($\mu = U/2$) written in a particle-hole symmetric form

$$H = -t \sum_{\langle i,j \rangle \sigma} \left( c_{i\sigma}^\dagger c_{j\sigma} + \text{h.c} \right) + \frac{U}{2} \sum_i (n_i - 1)^2 - \frac{U}{2} N_s, \qquad (3.26)$$

where $N_s$ is the number of lattice sites and $N_s \to \infty$ is implicitly assumed.

**Slave-spin formulation**

On each lattice site we introduce an auxiliary pseudospin $\mathbf{I}$ with eigenstates

$$I^z |\pm\rangle = \pm \frac{1}{2} |\pm\rangle \qquad (3.27)$$

encoding doubly and empty occupied sites ($|+\rangle$) and singly occupied sites ($|-\rangle$) with excitation energies of the order of $U$. In addition, Fermi creation and destruction operators $f_\sigma^{(\dagger)}$ are introduced to describe the low-energy quasiparticle degrees of freedom. The physical creation (annihilation) operator of the original model is represented as

$$c_\sigma^{(\dagger)} = 2I^x f_\sigma^{(\dagger)}. \qquad (3.28)$$

The physical states in the enlarged local Hilbert-space are

$$|e\rangle = |+\rangle|0\rangle, \quad , |p_\sigma\rangle = |-\rangle|\sigma\rangle, \quad |d\rangle = |+\rangle|2\rangle, \qquad (3.29)$$

where $|0\rangle$ is the vacuum of the $f$-fermions,

$$|\sigma\rangle = f_\sigma^\dagger |0\rangle \quad \text{and} \quad |2\rangle = f_\uparrow^\dagger f_\downarrow^\dagger |0\rangle. \tag{3.30}$$

In the lattice system the above definitions are generalized for each lattice site $i$.

Let us define a local charge

$$Q_i := \left[ I_i^z + \frac{1}{2} - (n_i - 1)^2 \right]^2 = \frac{1}{2} + I_i^z \left[ 1 - 2(n_i - 1)^2 \right], \tag{3.31}$$

where $n_i = \sum_\sigma f_{i\sigma}^\dagger f_{i\sigma}$ and $Q_i$ has eigenvalues 0 and 1 corresponding to the local physical subspace $\mathcal{H}_i^{(0)}$ and the orthogonal complement $\mathcal{H}_i^{(1)}$. Thus, the local Hilbert space $\mathcal{H}_i$ is decomposed according to

$$\mathcal{H}_i = \mathcal{H}_i^{(0)} \oplus \mathcal{H}_i^{(1)}, \quad \mathcal{H}_i^{(q)} = \{|\psi\rangle \in \mathcal{H}_i; Q_i|\psi\rangle = q|\psi\rangle\}, \quad q = 0, 1. \tag{3.32}$$

The projection onto the full physical subspace is achieved by imposing for each lattice site the constraint

$$Q_i = 0, \quad \forall i. \tag{3.33}$$

As a result, we can write the Hubbard interaction in the physical subspace solely using pseudospin operators $I_i^z$ and the original Hamiltonian is represented as

$$H' = -4t \sum_{\langle i,j\rangle,\sigma} I_i^x I_j^x \left( f_{i\sigma}^\dagger f_{j\sigma} + \text{h.c.} \right) + \frac{U}{2} \sum_i I_i^z - \frac{U}{4} N_s. \tag{3.34}$$

As long as the constraint (3.33) is fulfilled, the Hamiltonian (3.34) is equivalent to the original model (3.26).

**Conserved charge and local U(1)-gauge symmetry**

Because the physical subspace is invariant under the dynamics generated by $H'$, the generalized local charge $Q_i$ is a conserved quantity and therefore *has to* commute with the Hamiltonian

$$[Q_i, H'] = 0. \tag{3.35}$$

Equivalently, we consider the local unitary transformation

$$U_i(\varphi) = e^{i\varphi(Q_i - \frac{1}{2})} = e^{i\varphi I_i^z \left[ 1 - 2(n_i-1)^2 \right]} \tag{3.36}$$

## 3.3 A slave-spin theory for the particle-hole symmetric Hubbard model

generated by $Q_i$ and show that $U_i(\varphi)$ defines a continuous local symmetry of $H'$:

$$\tilde{H}' \equiv U_i(\varphi) H' U_i^\dagger(\varphi) = H'. \tag{3.37}$$

Since operators acting on sites different than $i$ are not affected by the above transformation and because obviously $U_i(\varphi) I_i^z U_i^\dagger(\varphi) = I_i^z$ it is sufficient to prove that the physical creation and destruction operators are left invariant under the operation defined by $U_i(\varphi)$. That this is in fact the case is most easily seen by writing them in the local occupation-number basis

$$2I_i^x f_{i\sigma} = \underbrace{|+\rangle\langle-| \otimes |0\rangle\langle\sigma| + |-\rangle\langle+| \otimes |\bar{\sigma}\rangle\langle 2|}_{(1-Q_i)2I_i^x f_{i\sigma}}$$
$$+ \underbrace{|-\rangle\langle+| \otimes |0\rangle\langle\sigma| + |+\rangle\langle-| \otimes |\bar{\sigma}\rangle\langle 2|}_{Q_i 2I_i^x f_{i\sigma}},$$

$$2I_i^x f_{i\sigma}^\dagger = \underbrace{|-\rangle\langle+| \otimes |\sigma\rangle\langle 0| + |+\rangle\langle-| \otimes |2\rangle\langle\bar{\sigma}|}_{(1-Q_i)2I_i^x f_{i\sigma}^\dagger}$$
$$+ \underbrace{|+\rangle\langle-| \otimes |\sigma\rangle\langle 0| + |-\rangle\langle+| \otimes |2\rangle\langle\bar{\sigma}|}_{Q_i 2I_i^x f_{i\sigma}^\dagger},$$

such that we can convince our-self of the invariance

$$U_i(\varphi) 2I_i^x f_{i\sigma}^{(\dagger)} U_i^\dagger(\varphi) = 2I_i^x f_{i\sigma}^{(\dagger)}. \tag{3.38}$$

This proves the conservation of the local charge $Q_i$. The symmetry $U_i(\varphi)$ is called a local U(1)-gauge symmetry and is a consequence of the enlarged Hilbert space. We can now write the partition function as

$$\mathcal{Z} = \text{Tr}\left[e^{-\beta H'} \prod_i (1-Q_i)\right] \tag{3.39}$$

where the trace is over the full Hilbert space involving both physical and non-physical states.

### 3.3.2 Mean-field theory

Within the mean-field solution, we assume product states in pseudospin and fermion degrees of freedom, thereby fully relaxing the constraint (3.33). The rational behind this treatment is the fact that we can approximately distinguish two time scales. Indeed, as shown below, the dynamics of the pseudospins is roughly governed by $\hbar/\max(U, U_c)$ whereas that of the pseudo-fermions is determined by $\sim \hbar/t$. Although this observation justifies to some extend the mean-field decoupling in retrospect, it should be considered as a first step on which a more sophisticated analysis can be based.

It is noteworthy, however, that the canonical anti-commutation relations of the physical destruction and creation operators are preserved on average,

$$\langle\{c_{i\sigma}, c_{j\sigma'}^\dagger\}\rangle = 4\langle I_i^x I_j^x\rangle \langle\{f_{i\sigma}, f_{j\sigma'}^\dagger\}\rangle = \delta_{ij}\delta_{\sigma\sigma'}, \quad (3.40)$$

where $\langle\ldots\rangle$ denotes the average over mean-field eigenstates. As a consequence, the single-particle spectral weight is correctly normalized as long as the spin identity

$$(I_i^x)^2 = \frac{1}{4} \quad (3.41)$$

is respected. As a result of the mean-field decoupling we obtain two effective Hamiltonians: The fermion problem assumes the form of a non-interacting tight-binding Hamiltonian with the hopping amplitude renormalized by a factor

$$g_{ij} = 4\langle I_i^x I_j^x\rangle, \quad (3.42)$$

where $i, j$ are nearest neighbors, i.e.,

$$H_f = -t\sum_{\langle i,j\rangle,\sigma} g_{ij}\left(f_{i\sigma}^\dagger f_{j\sigma} + \text{h.c.}\right). \quad (3.43)$$

On the other hand, the pseudospin problem reduces to the transverse-field Ising model

$$H_I = -\sum_{\langle i,j\rangle} J_{ij} I_i^x I_j^x + h\sum_i I_i^z \quad (3.44)$$

with the transverse field $h = U/2$ and the exchange coupling

$$J_{ij} = 4t\sum_\sigma \left(\langle f_{i\sigma}^\dagger f_{j\sigma}\rangle + \text{c.c}\right). \quad (3.45)$$

The transverse-field Ising model (3.44) is a prime example of a system displaying a quantum critical point at a critical ratio $(2h/J)_c = (U/J)_c$, separating a

## 3.3 A slave-spin theory for the particle-hole symmetric Hubbard model

magnetically ordered from a quantum paramagnet. Equations (3.42) and (3.45) are the two coupled self-consistency equations to be solved in the mean-field approximation. We note that there is always the trivial solution $g_{ij} = J_{ij} = 0$ of these equations which describes the system in the atomic limit.

In the following we work in the zero temperature limit and we restrict our analysis to translation-invariant states for which

$$J_{ij} = J = -\frac{16}{z} \int_{-zt}^{\varepsilon_F} d\varepsilon \varepsilon \rho_\sigma(\varepsilon) \approx 2.67 t. \tag{3.46}$$

Here, $\rho_\sigma(\varepsilon)$ is the non-interacting density of states per spin, $\varepsilon_F = 0$ is the Fermi energy for the pseudo fermions and $z$ is the coordination number. The integral is evaluated for a cubic lattice in three dimensions.

### 3.3.3 Brinkman-Rice transition

We start our discussion of the transverse-field Ising model (3.44) by applying a single-site molecular-field approximation. Noteworthy, on this level of the approximation, the mean-field self-consistency (3.42) and (3.45) leads to the Brinkman-Rice transition known from the Gutzwiller approximation, see (3.22). Let us introduce the mean magnetization $\langle I_0^x \rangle$ and

$$H_I^{MF} = \tilde{h} \sum_i \bar{I}_i^z, \quad \tilde{h} = \sqrt{h^2 + (Jz\langle I_0^x \rangle)^2}, \tag{3.47}$$

where the pseudospin has been rotated due to the action of the molecular field $Jz\langle I_0^x \rangle$,

$$\bar{\mathbf{I}}_i = e^{i\alpha I_i^y} \mathbf{I} e^{-i\alpha I_i^y}, \quad \tan \alpha = \frac{Jz\langle I_0^x \rangle}{h}. \tag{3.48}$$

Self-consistency of $\langle I_0^x \rangle$ yields the pseudospin magnetization and from (3.42) we obtain the hopping renormalization factor as follows:

$$g = 4\langle I_0^x \rangle^2 = 1 - \left(\frac{2h}{Jz}\right)^2 = 1 - u^2, \quad \text{for } u \leq 1, \quad u = \frac{U}{U_c}, \quad U_c = Jz, \tag{3.49}$$

and $g = 0$ for $u > 1$. In particular, on this level of the approximation, the effective mass $m^*/m = 1/g$ diverges at the critical interaction strength $U_c$, indicating the transition to the localized state.

### 3.3.4 Schwinger bosons and pseudospins

The connection to the slave-boson methods is established by use of the Schwinger-boson representation (3.2) of the pseudospin operators, which in addition allows to take into account the effect of quantum fluctuations and yields the low-lying excitations. Thus, we introduce two sets of Bose creation and annihilation operators $x_i^{(\dagger)}$ and $y_i^{(\dagger)}$ to represent the spin algebra

$$I_i^+ = y_i^\dagger x_i, \quad I_i^- = x_i^\dagger y_i, \quad I_i^z = y_i^\dagger y_i - \frac{1}{2}, \quad x_i^\dagger x_i + y_i^\dagger y_i = 1. \qquad (3.50)$$

In the Schwinger-boson formulation, the pseudospin rotation given in (3.48) translates to an unitary transformation of the Bose creation and destruction operators

$$\begin{pmatrix} x_i \\ y_i \end{pmatrix} = \begin{pmatrix} \cos\frac{\alpha}{2} & -\sin\frac{\alpha}{2} \\ \sin\frac{\alpha}{2} & \cos\frac{\alpha}{2} \end{pmatrix} \begin{pmatrix} a_i \\ b_i \end{pmatrix}$$

$$= \begin{pmatrix} \sqrt{1-2d^2} & -\sqrt{2}d \\ \sqrt{2}d & \sqrt{1-2d^2} \end{pmatrix} \begin{pmatrix} a_i \\ b_i \end{pmatrix}, \qquad (3.51)$$

where the $a$- and $b$-bosons can be interpreted as the Schwinger bosons of the rotated pseudospin $\bar{I}_i$. In the following, we denote the canonically transformed Ising model by $H_B(d)$ with the parameter $d$ specifying the transformation (3.51). More details are given in App. A.1.

#### Classical ground state

The result of the molecular-field approximation is recovered by assuming a product form of the wave-function

$$|B\rangle = \prod_i a_i^\dagger |0\rangle, \qquad (3.52)$$

and optimizing the energy

$$E(d) = \langle B|H_B(d)|B\rangle \qquad (3.53)$$

with respect to the parameter $d$. We recover again the Gutzwiller result (3.22). The state (3.52) is the classical ground-state of the transverse-field Ising model.

*3.3 A slave-spin theory for the particle-hole symmetric Hubbard model*

### 3.3.5 Role of fluctuations in three dimensions

The formalism developed so far offers a framework to study the excitation spectrum of the transverse-field Ising model. We show that these excitations lead to the incoherent one-particle excitations of the Hubbard model (upper and lower Hubbard bands). Furthermore, quantum fluctuations renormalize the classical ground state obtained in the molecular-field approximation. In general, the role of fluctuations crucially depends on the dimensionality of the system. Here we restrict our analysis to the three dimensional (cubic) lattice where fluctuations around the classical ground state are small for most parameters. Nevertheless, the finite dimensionality is reintroduced in our analysis which causes interesting changes in the nature of the Mott transition as compared to the infinite dimensional result. Formally, we derive an effective Hamiltonian describing excitations around the renormalized classical ground state and use a Bogoliubov transformation to diagonalize it. Details of the calculation are presented in App. A.1. In the following we discuss the results.

**Pseudo-spin-wave mode and Mott-Hubbard gap**

The (low-lying) excitations of the transverse-field Ising model are described by the gapped pseudospin-wave spectrum

$$\hbar\omega_{\mathbf{k}} = \frac{U_c}{2} \times \begin{cases} \sqrt{1 + u^2 \gamma_{\mathbf{k}}}, & \text{for } u \leq 1, \\ \sqrt{u^2 + u\gamma_{\mathbf{k}}}, & \text{for } u > 1, \end{cases} \quad (3.54)$$

with $\gamma_{\mathbf{k}} = -(1/3) \sum_{i=1}^{3} \cos k_i$ and an excitation gap $\Delta = \hbar\omega(0)$. The quantum criticality at $u = 1$ is reflected in the softening of the mode (3.54). For $u > 1$, the jump $\Delta\mu$ in the chemical potential from hole to particle doping amounts to twice the excitation gap,

$$\Delta\mu = 2\Delta = U\sqrt{1 - \frac{1}{u}}; \quad (3.55)$$

the above pseudospin-mode corresponds to the gapped charge excitation of the Mott insulator and (3.55) coincides with the expression found for the Mott-Hubbard gap in the Kotliar-Ruckenstein formulation [98, 100]. The band of the pseudospin mode (3.54) is shown in Fig. 3.3(a) for different values of the interaction strength $u$.

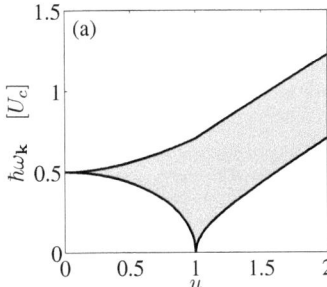 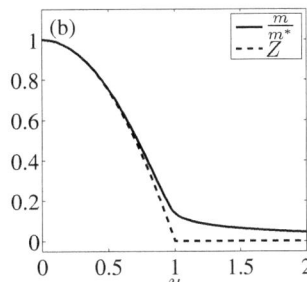

Figure 3.3: (a) The pseudospin excitation spectrum obtained in the spin-wave analysis of the transverse-field Ising model in three dimensions as function of $u = U/U_c$. (b) The inverse effective mass $m/m^*$ and the quasiparticle weight $Z$ as function of $U/U_c$. Note that at the Mott transition $u = 1$, $Z$ vanishes in contrast to $m/m^*$.

### Renormalized ground state

Quantum fluctuations lead to a renormalization of the ground-state energy of the model (3.26). We find

$$\begin{aligned}
E_G(u \leq 1) + \frac{UN_s}{2} &= -\frac{U_c N_s}{4}\left[\frac{(1-u)^2}{2} + \frac{1}{N_s}\sum_{\mathbf{k}}\left(1 - \sqrt{1+u^2\gamma_{\mathbf{k}}}\right)\right], \\
E_G(u > 1) + \frac{UN_s}{2} &= -\frac{UN_s}{4}\frac{1}{N_s}\sum_{\mathbf{k}}\left(1 - \sqrt{1+\gamma_{\mathbf{k}}/u}\right),
\end{aligned} \qquad (3.56)$$

where the terms proportional to $\sum_{\mathbf{k}}(\ldots)$ represent quantum corrections to the result of the Gutzwiller approximation. Note that $E_G < 0$ for any finite $u$, meaning that the mean-field solution with $g = J = 0$ is always higher in energy for any finite $u$. This is in contrast to the single-site molecular field approximation where the solution for $u > 1$ is trivial.

### Effective mass and quasiparticle weight in three dimensions

As in previous studies [101], we obtain that fluctuations reintroduce inter-site correlations beyond the molecular-field value. In particular, there is a distinction

## 3.3 A slave-spin theory for the particle-hole symmetric Hubbard model

between the quasiparticle weight and the effective mass renormalization of the quasiparticles, see Fig. 3.3(b). While the hopping renormalization factor $g = m/m^*$ stays finite across the Mott transition, the quasiparticle weight $Z$ still vanishes for $u \to 1$. This distinction is apparent in the pseudospin representation where the two quantities are expressed as

$$Z = 4\langle I_i^x \rangle^2 \quad \text{and} \quad \frac{m}{m^*} = g = 4\langle I_i^x I_j^x \rangle \quad \text{with} \quad i,j \quad \text{n.n.} \tag{3.57}$$

More general, the electronic self-energy obtains a **k** dependence which, in the metallic phase at particle-hole symmetry, is of the form

$$\Sigma(\omega, \mathbf{k}) = (1 + Z^{-1})\omega + (\frac{g}{Z} - 1)\varepsilon_\mathbf{k} \tag{3.58}$$

for **k** near the Fermi surface and for small $\omega$. In the insulating phase, the effective mass is given by

$$\frac{m}{m^*} = -\int d\varepsilon\, \rho_\sigma(\varepsilon) \frac{\varepsilon/D}{\sqrt{1 + \frac{1}{u}\varepsilon/D}} \approx \frac{U_c}{2U} \underbrace{\int d\varepsilon\, \rho_\sigma(\varepsilon)\, (\varepsilon/D)^2}_{=t/D=1/6} = \frac{U_c}{12U} \tag{3.59}$$

where $D = 6t$ is half of the band width. Thus, the spin-wave analysis captures correctly the energy scale $m/m^* \sim J/t$ for large $U$ with $J = \frac{4t^2}{U}$. The Fermi surface in the insulating phase is formed by spinons and is due to the fact that we have completely suppressed possible magnetic order.

**Spectral one-particle densiy**

In order to obtain the spectral one-particle density we use the Lehmann representation

$$A_\sigma(\omega) = \sum_n \left[ \left|\langle 0|c_{0\sigma}|n\rangle\right|^2 \delta(\omega - \omega_{n0}) + \left|\langle 0|c_{0\sigma}^\dagger|n\rangle\right|^2 \delta(\omega + \omega_{n0}) \right]. \tag{3.60}$$

where $|n\rangle$ denotes an eigenstate of the full Hamiltonian with energy $E_n$ and $\omega_{nm} = E_n - E_m$. In the slave-spin method used here the true eigenstates are approximated by the mean-field eigenstates obtained in the spin-wave analysis. We than have to calculate matrix elements of the form $\langle 0|I_0^x f_{\mathbf{q}\sigma}|n\rangle$. Details of the calculation are presented in the Appendix A.1.4. The spectral weight contains a

*Slave-particle methods for strong correlation physics*

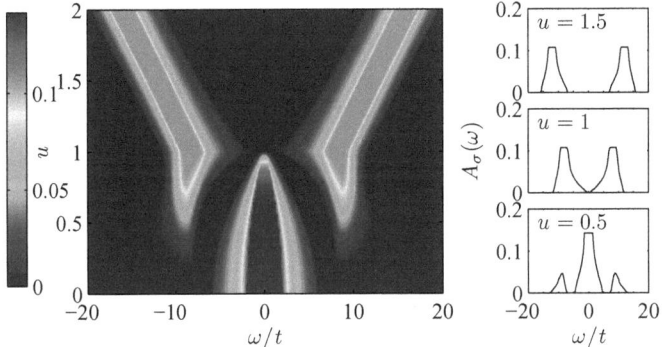

Figure 3.4: The contour plot shows the one-particle spectral density $A_\sigma(\omega)$ with the coherent Gutzwiller and the (preformed) Hubbard bands as function of energy $\omega/t$ and interaction strength $u$. On the right-hand side, $A_\sigma(\omega)$ is shown for fixed values of $u$ in the metallic and the insulating phase.

coherent quasiparticle peak $A_\sigma^{\text{coh}}(\omega)$ as well as an incoherent contribution $A_\sigma^{\text{inc}}(\omega)$. We find that the coherent contribution is given by

$$A_\sigma^{\text{coh}}(\omega) = \frac{Z}{g}\rho_\sigma(\omega/g), \quad g = \frac{m}{m^*}. \quad (3.61)$$

Note that $A_\sigma(0) = A_\sigma^{\text{coh}}(0) \propto Z/g$ gradually vanishes when approaching the Mott insulator, in contrast to the infinite dimensional result. In the metallic phase $u \leq 1$, the incoherent contribution is dominated by

$$A_\sigma^{\text{inc}}(\omega) \approx \frac{4D}{U_c} \int_0^\infty d\varepsilon\, \rho_\sigma(\varepsilon)$$
$$\times \begin{cases} \rho_\sigma\left[\frac{D}{u^2}\left(\frac{4(\omega-g\varepsilon)^2}{U_c^2} - 1\right)\right], & \Delta_- < \omega < \Delta_+ + gD; \\ \rho_\sigma\left[\frac{D}{u^2}\left(\frac{4(\omega+g\varepsilon)^2}{U_c^2} - 1\right)\right], & -\Delta_+ - gD < \omega < \Delta_-; \\ 0, & \text{else.} \end{cases} \quad (3.62)$$

43

## 3.3 A slave-spin theory for the particle-hole symmetric Hubbard model

where $\Delta_\pm = \frac{U_c}{2}\sqrt{1 \pm u^2}$ denote the edges of the excitation spectrum. In the insulating phase $u > 1$, $A_\sigma^{\text{coh}}(\omega) = 0$ and we find

$$A_\sigma(\omega) = A_\sigma^{\text{inc}}(\omega) = \frac{4D}{U_c} \int_0^\infty d\varepsilon\, \rho_\sigma(\varepsilon)$$

$$\times \begin{cases} \rho_\sigma\left[\frac{D}{u}\left(\frac{4(\omega-g\varepsilon)^2}{U_c^2} - u^2\right)\right], & \Delta_- < \omega < \Delta_+ + gD; \\ \rho_\sigma\left[\frac{D}{u}\left(\frac{4(\omega+g\varepsilon)^2}{U_c^2} - u^2\right)\right], & -\Delta_+ - gD < \omega < \Delta_-; \\ 0, & \text{else.} \end{cases} \quad (3.63)$$

The spectral density is shown in Fig. 3.4 for different values of the interaction strength. The gapped mode (3.54) found in the transverse-field Ising model leads to the incoherent weight around $\pm \max(U_c, U)/2$ in the spectral density [98]. In the metallic phase, we find the characteristic three peak structure with preformed Hubbard bands centered at $\hbar\omega \approx \pm U_c/2$ and a coherent Gutzwiller band at $\hbar\omega \approx 0$. The Gutzwiller band disappears at $u = 1$ and the Hubbard bands touch at $\hbar\omega = 0$. Using the expansion (3.59) of the effective mass for $u \gg 1$ and expression (3.63) we find that in the large $U$ limit the Hubbard bands assume a constant width of $U_c/2 \approx 8t$ and are separated by $U$. This band narrowing is in qualitative agreement with the retractable path approximation for a single hole doped into an infinite-$U$ Mott insulator [102].

**One-particle sum rule and fluctuation regime**

To estimate the validity of the molecular-field plus spin-wave calculation, we compare the magnitude of the fluctuations with the magnitude of the order parameter. We define $u^{\text{fl}}$ as the value of $u$ where both quantities are equal and define the fluctuation regime as the symmetric region around the critical point, $u = 1 \pm (1 - u^{\text{fl}})$. The magnitude of the fluctuations is measured by averaging $\langle \delta I_0^x \delta I_i^x \rangle$ over a volume containing the six nearest neighboring sites of site 0. The condition that the fluctuations are equal to the magnitude of the order parameter for some value $u^{\text{fl}}$ of the interaction strength then reads

$$\frac{1}{1+6}\left[\sum_{\langle 0,i \rangle} \langle \delta I_0^x \delta I_i^x \rangle + \langle (\delta I_0^x)^2 \rangle\right] = \langle I_0^x \rangle^2 \bigg|_{u^{\text{fl}}}. \quad (3.64)$$

For the cubic lattice considered here we obtain $u^{\text{fl}} \approx 0.9$. Hence, for $u \approx 1 \pm 0.1$ the fluctuation induced corrections to the molecular-field result are important and the validity of the above analysis is limited. In particular, the approximative nature of our treatment of the pseudospin problem violates $(I_i^x)^2 = 1/4$ and the spectral weight fails to be properly normalized. As shown in Fig. 3.5, the one-particle sum rule

$$\int d\omega A_\sigma(\omega) = 1 \qquad (3.65)$$

is fulfilled within 10-12%. The failure only manifests itself in the fluctuation regime, however.

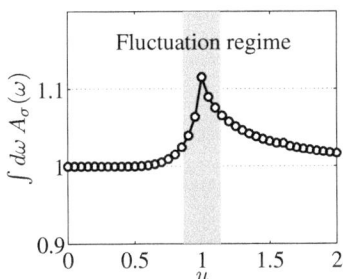

Figure 3.5: The integrated spectral weight of the one-particle spectral density in the pseudospin-wave analysis. The shaded region denotes the fluctuation regime where the magnitude of the fluctuations is of the same order than the order parameter.

## 3.4 The SBA approximation and DMFT

In recent years, DMFT [80] has become an important and widely used approximation for studying strongly correlated electron systems. This powerful scheme relies on the mapping of the lattice problem onto an effective single impurity Anderson model with self-consistently determined properties of the conduction

## 3.4 The SBA approximation and DMFT

electrons and is exact in infinite dimensions. In the metallic regime at low temperatures, the physical picture provided by the DMFT is very similar to the one of Brinkman and Rice using the Gutzwiller approximation [66] or the slave-boson mean-field approximation of Kotliar and Ruckenstein [64]. The goal of this chapter is to provide further insights into their mutual relation. In particular, we show that the SBA solution of the Hubbard model is equivalent to the DMFT approximation using the SBA to solve the auxiliary impurity problem. This equivalence in the formulation was for example exploit in the context of strongly correlated disordered systems in order to obtain a numerically more stable procedure [103, 104].

### 3.4.1 The dynamical mean-field approach

The key quantity on which DMFT focuses is the local Green's function at a given lattice site $i$:

$$G_{ii,\sigma}(\tau - \tau') \equiv -\langle T_\tau c_{i\sigma}(\tau) c_{i\sigma}^\dagger(\tau') \rangle. \tag{3.66}$$

The local Green's function (3.66) can be represented as that of a single atom (impurity orbital) coupled to an effective bath, as can be described by the Hamiltonian of an Anderson impurity model

$$H_{\text{AM}} = \sum_{l\sigma} \varepsilon_l a_{l\sigma}^\dagger a_{l\sigma} + \sum_{l\sigma} V_l \left( a_{l\sigma}^\dagger c_\sigma + c_\sigma^\dagger a_{l\sigma} \right) - \mu n + U n_\uparrow n_\downarrow \tag{3.67}$$

where $n = n_\uparrow + n_\downarrow$ and $n_\sigma = c_\sigma^\dagger c_\sigma$. The Fermi operators $a_{l\sigma}^{(\dagger)}$ describe the bath degrees of freedom ("the conduction electrons") with eigenmodes $\varepsilon_l$. They are coupled to the correlated impurity site through the hybridization matrix elements $V_l$. Representing the partition function as a path integral over Grassmann variables and integrating out the bath degrees of freedom (the $a$-fermions), we arrive at

$$\mathcal{Z}_{\text{imp}} = \int \mathcal{D}(c^+, c) e^{-S_{\text{imp}}[c^+, c]} \tag{3.68}$$

with

$$S_{\text{imp}}[c^+, c] = \sum_{\omega_n \sigma} c_\sigma^+(i\omega_n) \left[ -i\omega_n - \mu + \Delta(i\omega_n) \right] c_\sigma(i\omega_n)$$
$$+ U \int_0^\beta d\tau\, n_\uparrow(\tau) n_\downarrow(\tau), \tag{3.69}$$

where we have used the notation

$$c_\sigma(i\omega_n) = \int_0^\beta d\tau\, c_\sigma(\tau) e^{i\omega_n \tau}, \quad c_\sigma^+(i\omega_n) = \int_0^\beta d\tau\, c_\sigma^+(\tau) e^{-i\omega_n \tau}, \qquad (3.70)$$

with fermionic Matsubara frequencies $\omega_n = (2n+1)\pi/\beta$. The hybridization function $\Delta(i\omega_n)$ is given in terms of the bath parameters by

$$\Delta(i\omega_n) = \sum_l \frac{|V_l|^2}{i\omega_n - \epsilon_l}. \qquad (3.71)$$

Note that (3.71) (i.e. the parameters $V_l$ and $\epsilon_l$) is not known a priori but has to be found self-consistently, see below. The self-energy of the effective impurity model can be defined from the interacting local Green's function $G(i\omega_n)$ as

$$\Sigma_{\text{imp}}(i\omega_n) = i\omega_n + \mu - \Delta(i\omega_n) - G^{-1}(i\omega_n). \qquad (3.72)$$

On the other hand, the Green's function of the lattice model is given by

$$G(\mathbf{k}, i\omega_n) = \frac{1}{i\omega_n + \mu - \varepsilon_\mathbf{k} - \Sigma(\mathbf{k}, i\omega_n)} \qquad (3.73)$$

in which $\varepsilon_\mathbf{k}$ is the non-interacting dispersion and $\Sigma(\mathbf{k}, i\omega_n)$ is the lattice self-energy. The basic assumption in the DMFT is now that the lattice self-energy is purely local and can thus be represented by the self-energy of an adequately chosen impurity model

$$\Sigma(\mathbf{k}, i\omega_n) \approx \Sigma_{\text{imp}}(i\omega_n). \qquad (3.74)$$

In order to obtain the local part of the lattice Green's function we sum (3.73) over $\mathbf{k}$, and using (3.72) and (3.74), we arrive at the self-consistency condition

$$\frac{1}{N_s} \sum_\mathbf{k} \frac{1}{\Delta(i\omega_n) + G(i\omega_n)^{-1} - \varepsilon_\mathbf{k}} = G(i\omega_n). \qquad (3.75)$$

This self-consistency condition relates the dynamical mean-field $\Delta(i\omega_n)$ to the local Green's function $G(i\omega_n)$ of the effective impurity model (3.69). Therefore, the two functions $\Delta$ and $G$ are in principle fully determined. In practice, it is convenient to use an iterative procedure to solve (3.75): start with an initial choice of $\Delta(i\omega_n)$. Compute the local Green's function using an adequate *impurity solver*. Use the self-consistency condition (3.75) to obtain the new dynamical field $\Delta$. Iterate until convergence is achieved.

## 3.4 The SBA approximation and DMFT

### 3.4.2 SBA impurity solver

The DMFT scheme requires the (numerical or approximative) solution of the single impurity Anderson model for a given set of bath parameters. Usually, this is the most time-consuming part. It is therefore of great interest to have efficient and reliable impurity solvers, such as the continuous time quantum Monte Carlo algorithm [105] or the numerical renormalization group method [106]. However, when dealing with more complex situations such as disordered [103, 104] or other spatially non-uniform systems [39] as well as in combination with realistic band structure calculations [46] it is useful to have numerically cheap (but approximative) impurity solvers in order to scan a wide range of parameters. In this sense, we use in the following the slave-boson scheme of Kotliar and Ruckenstein as impurity solver for the low-temperature Fermi liquid regime. Comparison with quantum Monte Carlo data yields acceptable accuracy [107]. Furthermore, the parameters of the slave-boson approximation can be used for interpolating the self-energy over the whole frequency range [107]. Related methods are based on a slave-rotor representation [108] or on the Gutzwiller approximation scheme [109] to solve the auxiliary impurity problem.

**A simple impurity solver**

Applying the SBA to the single impurity Anderson model (3.67), the effective impurity action (3.69) is replaced by

$$\tilde{S}_{\text{imp}} = \sum_{\omega_n \sigma} f_\sigma^+(i\omega_n) \left[ -i\omega_n + \lambda + Z\Delta(i\omega_n) \right] f_\sigma(i\omega_n) + \beta \left( U d^2 - \lambda n - \mu n \right)$$

where $Z = z(n,d)^2$ is a function of the parameters $d$ and $n$, see Eq. (3.21). The impurity free energy

$$\Omega_{\text{imp}}(\beta, \mu) = -\frac{1}{\beta} \log \mathcal{Z}_{\text{imp}} \tag{3.76}$$

in the mean-field approximation is then given by

$$\tilde{\Omega}_{\text{imp}} = -\frac{2}{\beta} \sum_n \log \left\{ \beta \left[ -i\omega_n + \lambda + Z\Delta(i\omega_n) \right] \right\} + U d^2 - \lambda n - \mu n. \tag{3.77}$$

The parameters $d$, $n$ and $\lambda$ are determined by the saddle-point equations [103]

$$\frac{\partial Z}{\partial d} \frac{1}{\beta} \sum_{\omega_n} \Delta(i\omega_n) G^{\text{coh}}(i\omega_n) = -ZUd \tag{3.78}$$

$$\frac{\partial Z}{\partial n} \frac{1}{\beta} \sum_{\omega_n} \Delta(i\omega_n) G^{\text{coh}}(i\omega_n) = Z(\lambda + \mu) \tag{3.79}$$

$$\frac{1}{\beta} \sum_{\omega_n} G^{\text{coh}}(i\omega_n) = \frac{1}{2} Z n \tag{3.80}$$

where we have introduced the coherent part of the local Green's function

$$G^{\text{coh}}(i\omega_n) = \frac{Z}{i\omega_n - \lambda - Z\Delta(i\omega_n)}. \tag{3.81}$$

Using (3.72) the self-energy is thus given as

$$\Sigma_{\text{imp}}^{\text{KR}}(i\omega_n) = \frac{\lambda}{Z} + \left(1 - \frac{1}{Z}\right) i\omega_n \tag{3.82}$$

which can be viewed as an expansion for small frequencies.

### Equivalence between lattice SBA and DMFT + SBA

Using the SBA scheme as impurity solver in combination with the DMFT self-consistency relation (3.75) for the coherent part of the Green's function is in fact equivalent to the SBA of the lattice model. The validity of this statement is shown in the following: the self-concsistency relation reads

$$G^{\text{coh}}(i\omega_n) = \frac{1}{N_s} \sum_{\mathbf{k}} G^{\text{coh}}(\mathbf{k}, i\omega_n)$$

$$\frac{Z}{i\omega_n - \lambda - Z\Delta(i\omega_n)} = \int d\varepsilon \frac{Z\rho(\varepsilon)}{i\omega_n - \lambda - Z\varepsilon} = \tilde{D}\left(\frac{i\omega_n - \lambda}{Z}\right). \tag{3.83}$$

Here, we have used the Hilbert transform $\tilde{D}(z)$ of the non-interacting DOS $\rho(z)$ defined by

$$\tilde{D}(z) = \int d\varepsilon \frac{\rho(\varepsilon)}{z - \varepsilon}, \quad \rho(z) = \frac{1}{N_s} \sum_{\mathbf{k}} \delta(z - \varepsilon_{\mathbf{k}}). \tag{3.84}$$

## 3.4 The SBA approximation and DMFT

It follows that

$$\begin{aligned}\Delta(i\omega_n)G^{\text{coh}}(i\omega_n) &= \frac{i\omega_n - \lambda}{Z}\tilde{D}\left(\frac{i\omega_n - \lambda}{Z}\right) - 1 \\ &= \int d\varepsilon \frac{Z\varepsilon\rho(\varepsilon)}{i\omega_n - \lambda - Z\varepsilon} = \frac{1}{N_s}\sum_{\mathbf{k}}\varepsilon_{\mathbf{k}}G^{\text{coh}}(\mathbf{k}, i\omega_n)\end{aligned}$$

and consequently

$$\frac{1}{\beta}\sum_{\omega_n}\Delta(i\omega_n)G^{\text{coh}}(i\omega_n) = \frac{1}{N_s}\sum_{\mathbf{k}}Z\varepsilon_{\mathbf{k}}f_T(Z\varepsilon_{\mathbf{k}} + \lambda) \qquad (3.85)$$

which is the kinetic energy of the Hubbard model evaluated in the SBA. Thus, plugging the relation (3.85) back into the saddle-point equations (3.78)-(3.80) we recover the saddle-point equations obtained in the SBA of the Hubbard model.

# Chapter 4

# Strongly correlated electrons in a Hubbard heterostructure

> We present aspects of strong electron correlations in the Hubbard heterostructure introduced in Sec. 2.4.1. An effective Schrödinger-Poisson-Gutzwiller problem for the low-energy part is derived in the framework of the Kotliar and Ruckenstein slave-boson mean-field approximation (SBA) and solved self-consistently for a wide range of parameters. The paramagnetic solution in the strongly interacting limit contains coherent quasi-particles confined to a relatively narrow region near the interfaces which are responsible for metallic behavior. We discuss the novel electronic properties developing at the interface and interpret them as the fingerprints of an interfacial heavy-fermion state.

## 4.1 Introdcution

This chapter is largely based on the results presented in Ref. [69] which in turn was mainly inspired by the experiments on the $LaTiO_3/SrTiO_3$ interface reported by Ohtomo *et al.* [13] and the early theoretical works by Okamoto and Millis [28, 39].

Both constituents of the experimental system are insulating in bulk: the pure La-compound is a Mott insulator (MI) whereas the pure Sr-compound is a band

## 4.1 Introdcution

insulator (BI). Nevertheless, in atomically precise superlattices and heterojunctions, a metallic phase is stabilized. As pointed out earlier [28], the atomically precise fabrication, the near lattice match ($a \approx 3.9$ Å) and the chemical similarity of the two components offer a good starting point for theoretical studies and make it possible to study the influence of *electronic reconstruction*[1] alone. Indeed, there is strong experimental evidence [13, 110, 111] that in these homometallic heterostructures,[2] electronic charge transfer occurs as a mechanism to avoid the polar catastrophe.[3] As a result, mutual doping of band and Mott insulator leads to a quasi-two-dimensional electron gas confined to a relatively narrow region at the interface [13].

The LaTiO$_3$ bulk system has a Ti-$3d^1$ configuration and the SrTiO$_3$ compound has a Ti-$3d^0$ configuration. In this sense, the La ions embedded in the SrTiO$_3$ background define a "quantum well" confining the conduction electrons to the La region. In a single-particle picture, the bound states of this quantum well form quasi-two dimensional subbands of which several are partially filled. Nevertheless, the pure La compound is a Mott insulator and even if a Fermi liquid is formed in the artificially structured system (hence a single particle description is valid), it is an interesting question to which extend this simple picture has to be modified due to the presence of strong electron-electron interaction.

For the rest of this chapter we study a simplified model for heterostructures which are characterized by a BI/M/BI stacking. Here BI is a band-insulator such as SrTiO$_3$. By varying the onsite repulsion we can continuously change the properties of the sandwiched material M tuning it from a metal to a Mott insulator such as LaTiO$_3$. The model is based on a generalized single-band Hubbard model introduced in Sec. 2.4 of this thesis. We analyze the Fermi liquid state by means of the SBA introduced in Sec. 3.2 which allows us to give a systematic discussion on how strong onsite interaction in combination with a spatially nonuniform setup leads to differences in the electronic properties as compared to both the strongly correlated bulk systems or heterostructures where electron correlations play a minor role. In this way, we can identify novel electronic properties which are specific to the presence of the interface and strong

---

[1] See Sec. 2.2.2.
[2] Following Sec. 2.1.2 we denote a heterostructure of the type ABO$_3$/A'BO$_3$ as homometallic.
[3] See Sec. 2.2.1.

electronic correlations.

## 4.2 Quasiparticle description

The implicit assumption made in the SBA treatment of the Hubbard heterostructure is that the low-lying excitations can be characterized by a *local Fermi liquid*. This assumption will be stated more precisely in the following. Let us write the layer-dependent single-particle Green function in a mixed representation [22, 112]

$$G_{ll',\sigma}(\mathbf{k},\omega) = [\omega + \mu - \hat{t}(\mathbf{k}) - \hat{\Sigma}_\sigma(\mathbf{k},\omega)]^{-1}_{ll'}, \qquad (4.1)$$

where $\mathbf{k} = (k_x, k_y)$ and $l, l'$ subscript the layers. The matrix $\hat{t}(\mathbf{k})$ is the Fourier transformed hopping matrix

$$\hat{t}(\mathbf{k}) = \begin{pmatrix} \varepsilon_\mathbf{k} & -t & 0 & \dots \\ -t & \varepsilon_\mathbf{k} & -t & 0 \\ 0 & -t & \varepsilon_\mathbf{k} & -t \\ \dots & \dots & \dots & \dots \end{pmatrix}, \qquad (4.2)$$

with $\varepsilon_\mathbf{k}$ the free dispersion of the two-dimensional nearest-neighbor hopping tight-binding model given by

$$\varepsilon_\mathbf{k} = -2t(\cos k_x a + \cos k_y a). \qquad (4.3)$$

Furthermore, we have introduced the layer-dependent self-energy $\hat{\Sigma}_\sigma(\mathbf{k},\omega)$. In the following we consider the low-temperature limit $T \to 0$. The assumption of a local Fermi liquid means that the self-energy is local (thus independent of $\mathbf{k}$ and diagonal in the layer index[4]), and analytic for $\omega \to 0$. Hence, in this limit we write

$$\Sigma_{ll'}(\omega) = \delta_{ll'} \left[ \mu + \frac{\lambda_l}{z_l^2} + (1 - \frac{1}{z_l^2})\omega \right], \qquad (4.4)$$

where we have neglected the imaginary part, which is expected to be proportional to $\omega^2$ and have suppressed the spin index $\sigma$. The meaning of the different

---
[4]See also Sec. 3.4.1 about the dynamical mean-field theory (DMFT).

## 4.2 Quasiparticle description

parameters in Eq. (4.4) are clarified below. From Eq. (4.4) it follows that the layer-dependent wave-function renormalization factor,

$$\hat{Z} = \left[1 - \left.\frac{\partial \hat{\Sigma}(\omega)}{\partial \omega}\right|_{\omega=0}\right]^{-1}, \qquad (4.5)$$

is purely local,

$$Z_{ll'} = \delta_{ll'} z_l^2. \qquad (4.6)$$

$\lambda_l$ in Eq. (4.4) is an effective layer-dependent chemical potential including Hartree like (electrostatic) contributions from the long-range Coulomb potential as well as $\omega = 0$ contributions from the dynamical self-energy due to the onsite repulsion. The low-energy part of Green's function may be written in terms of an effective Hamiltonian

$$\hat{G}_\sigma(\mathbf{k},\omega) = \sqrt{\hat{Z}}\left[\omega - \hat{H}_{\text{eff}}(\mathbf{k})\right]^{-1}\sqrt{\hat{Z}} + \hat{G}_\sigma^{\text{inc}}(\mathbf{k},\omega), \qquad (4.7)$$

where the effective Hamiltonian is given by

$$\hat{H}_{\text{eff}}(\mathbf{k}) = \begin{pmatrix} z_1^2 \varepsilon_\mathbf{k} + \lambda_1 & -tz_1 z_2 & 0 & \cdots \\ -tz_2 z_1 & z_2^2 \varepsilon_\mathbf{k} + \lambda_2 & -tz_2 z_3 & 0 \\ 0 & -tz_3 z_2 & z_3^2 \varepsilon_\mathbf{k} + \lambda_3 & -tz_3 z_4 \\ \cdots & \cdots & \cdots & \cdots \end{pmatrix}, \qquad (4.8)$$

and $\hat{G}_\sigma^{\text{inc}}(\mathbf{k},\omega)$ denotes the incoherent part of Green's function, which we will neglect for $\omega \to 0$. Let us introduce the eigenstates $|\mathbf{k}\nu\rangle$ of $\hat{H}_{\text{eff}}(\mathbf{k})$ as

$$\hat{H}_{\text{eff}}(\mathbf{k})|\mathbf{k}\nu\rangle = E_{\mathbf{k}\nu}|\mathbf{k}\nu\rangle, \qquad (4.9)$$

where $\nu$ is the subband index and the envelope wave function is given by $\psi_{\mathbf{k}\nu}(l) = \langle l|\mathbf{k}\nu\rangle$. The $\mathbf{k}$ dependence of these quantities enters through $\varepsilon_\mathbf{k}$,

$$E_{\mathbf{k}\nu} \equiv E_\nu(\varepsilon_\mathbf{k}), \quad \psi_{\mathbf{k}\nu} \equiv \psi_{\varepsilon_\mathbf{k}\nu}. \qquad (4.10)$$

Eventually, we find for $\omega \to 0$

$$\left[\hat{G}_\sigma(\mathbf{k},\omega)\right]_{ll'} = \sum_\nu \frac{z_l \psi_{\mathbf{k}\nu}(l)\psi_{\mathbf{k}\nu}(l') z_{l'}}{\omega - E_{\mathbf{k}\nu}}. \qquad (4.11)$$

For later use we also introduce the Green function of the quasiparticles,

$$\left[\hat{G}_\sigma^{\mathrm{QP}}(\mathbf{k},\omega)\right]_{ll'} = \sum_\nu \frac{\psi_{\mathbf{k}\nu}(l)\psi_{\mathbf{k}\nu}(l')}{\omega - E_{\mathbf{k}\nu}}, \quad (4.12)$$

which is diagonal in the quasiparticle subband basis. Furthermore, with (4.12) let us define the *local* quasiparticle density of states (DOS)

$$\rho_l^*(\omega) = -\frac{1}{N_\parallel \pi} \sum_{\mathbf{k}\sigma} \mathrm{Im}\, G_{ll,\sigma}^{\mathrm{QP}}(\mathbf{k},\omega+i0^+) = \frac{2}{N_\parallel} \sum_{\mathbf{k}\nu} \delta(E_{\mathbf{k}\nu}-\omega)|\psi_{\mathbf{k}\nu}(l)|^2 \quad (4.13)$$

as well as the *total* quasiparticle DOS

$$\rho^*(\omega) = \sum_l \rho_l^*(\omega) = 2 \sum_\nu Z_{\omega\nu}^{-1} \rho_\sigma(\varepsilon_{\omega\nu}). \quad (4.14)$$

Here, $N_\parallel$ denotes the number of lattice sites per layer. In (4.14) we have introduced the non-interacting DOS per spin projection

$$\rho_\sigma(\varepsilon) = \frac{1}{N_\parallel} \sum_{\mathbf{k}} \delta(\varepsilon_{\mathbf{k}}-\varepsilon), \quad (4.15)$$

$\varepsilon_{\omega\nu}$ is defined through the equation $E_\nu(\varepsilon_{\omega\nu}) = \omega$ and $Z_{\omega\nu} = \partial E_\nu(\varepsilon_{\omega\nu})/\partial\varepsilon$.

## 4.3 Slave-boson mean-field treatment

The SBA in its simplest form realizes the above scenario of a local Fermi liquid. Indeed, it offers a way to self-consistently determine the form (4.4) of the self-energy by solving a *Schrödinger-Poisson-Gutzwiller problem*. More specifically, three sets of variables have to be determined: (i) the effective quasiparticle potential $\lambda_l$, (ii) the electronic charge distribution $n_l$ and (iii) the amplitude $d_l$ of the double-occupancy density in layer $l$. The self-consistency equations for these variables are given below: (4.29)-(4.31). Beside the minimization of the free energy with respect to the double occupancy, their solution involves the diagonalization of the effective Hamiltonian (4.8) as well as the determination of the electrostatic potential by the Poisson equation. The schematics of the algorithm used to solve this problem is shown in Fig. 4.1 and more details are given in the subsequent paragraphs.

## 4.3 Slave-boson mean-field treatment

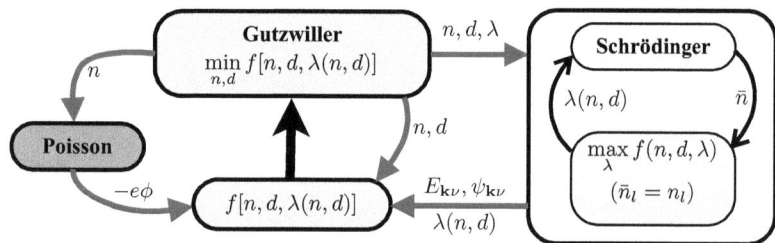

Figure 4.1: Schematic diagram of the algorithm used to solve the Schrödinger-Poisson-Gutzwiller problem obtained in the SBA treatment of the Hubbard heterostructure.

### 4.3.1 Ambiguity of Schrödinger-Poisson-Gutzwiller

There is an ambiguity in the above sketched approach which originates from the fact that the treatment of the local repulsion differs from the long-ranged (electrostatic) problem. Within the slave-boson method, this ambiguity manifests itself as follows: In the physical subspace the operator identity

$$U\hat{n}_{i\uparrow}\hat{n}_{i\downarrow} = (U - U_0)\hat{d}_i^\dagger \hat{d}_i + U_0 \hat{n}_{i\uparrow}\hat{n}_{i\downarrow} \tag{4.16}$$

holds with an arbitrary constant[5] $U_0$. However, the mean-field results depend on the choice of $U_0$ and a well-defined procedure to fix its value is necessary. Because usually the physically relevant parameter regime is $U \geq E_C$ we define a *minimal interacting* limit by $U = E_C$ and demand that in this limit the SBA approximation reproduces the Hartree mean-field approximation. Hence, we set $U_0 = E_C$ meaning that double occupancy is only significantly reduced for $U > E_C$. In our approach the relevant parameter describing the correlation induced deviation from the Hartree mean-field results is therefore

$$U_r = U - E_C. \tag{4.17}$$

---
[5]More general, one can introduce a site-dependent shift $U_0(i)$.

## 4.3.2 Effective one-dimensional Schrödinger equation

We use the convention $i = (j, l)$ where $l$ is the layer index ($z$-direction) and $j$ specifies the site in each individual layer ($x, y$-direction). We fix the origin $z = 0$ in the center of the sandwich and set the lattice spacing $a = 1$. In the following we shall assume a state which is invariant under translations in the planes. Then we have to determine the amplitude of doubly-occupied sites $d_l$ and the total charge distribution $n_l$ in layer $l$ as well as Lagrange multipliers $\lambda_l$ to enforce the self-consistency of the charge distribution. The SBA Hamiltonian can be diagonalized by a canonical transformation

$$f_{jl\sigma} = \frac{1}{\sqrt{N_\parallel}} \sum_{\mathbf{k}\nu} \psi_{\mathbf{k}\nu}(l) e^{i\mathbf{k}\cdot\mathbf{R}_j} f_{\mathbf{k}\nu\sigma}$$

which is composed of an in-plane Fourier transformation and of a transformation to the solutions of an effective one-dimensional Schrödinger equation of the form (4.8):

$$\left[(z_l^2 - 1)\varepsilon_\mathbf{k} + \lambda_l\right] \psi_{\mathbf{k}\nu}(l) - t \sum_{\gamma=\pm 1} z_l z_{l+\gamma} \psi_{\mathbf{k}\nu}(l+\gamma) = (E_{\mathbf{k}\nu} - \varepsilon_\mathbf{k}) \psi_{\mathbf{k}\nu}(l). \quad (4.18)$$

Here, $\varepsilon_\mathbf{k}$ is given in (4.3). The bound states of Eq. (4.18) define subbands of a quasi-two-dimensional metal. Furthermore, the symmetry $z \leftrightarrow -z$ allows to choose eigenstates with a fixed parity $(-1)^{\nu-1}$,

$$\psi_{\mathbf{k}\nu}(-l) = (-1)^{\nu-1} \psi_{\mathbf{k}\nu}(l). \quad (4.19)$$

The in-plane hopping is renormalized by a factor $z_l^2$ - corresponding to a layer dependent mass-renormalization $m_l^*/m = 1/z_l^2$ - and the hopping between different layers by $z_l z_{l'}$ where [64–66]

$$z_l(n_l, d_l) = \frac{\sqrt{(1 - n_l + d_l^2)(n_l - 2d_l^2)} + d_l\sqrt{n_l - 2d_l^2}}{\sqrt{n_l(1 - n_l/2)}}. \quad (4.20)$$

In Eq. (4.18), the inhomogeneous setup leads to an effective one-dimensional potential $(z_l^2 - 1)\varepsilon_\mathbf{k} + \lambda_l$ along the heterostructure which can vary essentially for different values of the transverse momentum $\mathbf{k}$, i.e., the in-plane kinetic energy. The diagonalization of Eq. (4.18) therefore has to be performed for each $\varepsilon_\mathbf{k}$ separately.

## 4.3 Slave-boson mean-field treatment

| $n$ | 0 | 1 | 2 |
|---|---|---|---|
| $\Delta(n)$ | -3.9003 | 0.0079 | $\|\ldots\| < 10^{-4}$ |
| $\Gamma(n+0.5)$ | -0.1388 | -0.0003 | $\|\ldots\| < 10^{-4}$ |

Table 4.1: Correction terms used for calculating the electrostatic energy.

### 4.3.3 Longe range Coulomb interaction

The resulting (screened) electrostatic potential energy in layer $l$,

$$-e\phi_l = V_l + \sum_{l'} W_{ll'} n_{l'}, \qquad (4.21)$$

is composed of the counter-ion potential $V_l$ and of the mean-field electronic contribution derived from the interaction matrix $W_{ll'}$ between electrons in layer $l$ and $l'$ of the heterostructure. In the end, $\phi_l$ has to be found self-consistently.

Solving the Poisson equation for a discrete set of point-like charges located at the respective lattice sites, we find the following formal expression[6] for the interaction matrix

$$\frac{W_{ll'}}{E_C} = \sum_{(nm)} \frac{(1-\delta_{ll'})}{\sqrt{n^2+m^2+(l-l')^2}} + \delta_{ll'} \left( \frac{U_0}{4E_C} + \sum_{(nm)\neq(00)} \frac{1}{\sqrt{n^2+m^2}} \right). \qquad (4.22)$$

The term proportional to $\frac{U_0}{4E_C}$ arises from the shift in $U$ discussed in Sec. 4.3.1. A convenient way to calculate the summation is to replace each layer by a uniformly charged plane and add numerically determined correction terms. We then have

$$\frac{W_{ll'}}{E_C} = \delta_{ll'} \frac{U_0}{4E_C} - 2\pi|l-l'| + \Delta(|l-l'|). \qquad (4.23)$$

The first few correction terms $\Delta(|l-l'|)$ are given in Table 4.1. As expected, they are significant very close to ($|l-l'|=1$) or within ($l=l'$) the considered layer [113]. The ion-potential is given by

$$\frac{V_l}{E_C} = -\sum_{(nm)} \sum_{k=1}^{N} \frac{1}{\sqrt{(n+\frac{1}{2})^2 + (m+\frac{1}{2})^2 + (l-l_k^{\text{ion}})^2}}, \qquad (4.24)$$

---

[6]The right-hand side of (4.22) should be understood as the regularized expression where the divergent contribution is canceled by the charge neutrality.

where $N$ is the number of counter-ion layers, $l_k^{\text{ion}}$ denotes the position of the $k$th ion-plane and $l - l_k^{\text{ion}}$ assumes half-integer values. In a similar way we find

$$\frac{V_l}{E_C} = \sum_{k=1}^{N} 2\pi |l - l_k^{\text{ion}}| + \Gamma(|l_k^{\text{ion}} - l|), \qquad (4.25)$$

and the most relevant correction terms $\Gamma(|l - l_k^{\text{ion}}|)$ are given in Tab. 4.1. When considering a superlattice with period $L$ instead of a quantum well geometry, one has to replace in (4.23) and (4.25) the elementary solution of a single charged plane, $2\pi E_C |l - l'|$, by the solution of a periodic array with the period $L$ of the superlattice

$$\frac{2\pi E_C}{L} |l - l'| (L - |l - l'|), \quad 0 \leq l, l' \leq L. \qquad (4.26)$$

More details can be found in [71].

### 4.3.4 Free energy and self-consistency

The fields $d_l$, $n_l$ and $\lambda_l$ are determined by the stationary point of the free energy density at the inverse temperature $\beta = 1/k_B T$,

$$\begin{aligned}
f(n, d, \lambda) &= -\frac{2}{\beta N_\|} \sum_{\mathbf{k}\nu} \ln \left[ 1 + e^{-\beta E_{\mathbf{k}\nu}(n,d,\lambda)} \right] \\
&\quad + U_r \sum_l d_l^2 + \frac{1}{2} \sum_{ll'} n_l W_{ll'} n_{l'} - \sum_l (\lambda_l - V_l) n_l,
\end{aligned} \qquad (4.27)$$

under the constraint of charge neutrality

$$\sum_l n_l = N. \qquad (4.28)$$

The saddle-point equations incorporating this condition are nonlinear, coupled self-consistency equations:

$$n_l = \bar{n}_l, \qquad (4.29)$$

$$d_l = -\frac{1}{U_r} \left( \frac{\partial z_l^2}{\partial d_l} \bar{\varepsilon}_l - t \sum_{\gamma=\pm 1} z_{l+\gamma} \frac{\partial z_l}{\partial d_l} \bar{\chi}_l^{(\gamma)} \right), \qquad (4.30)$$

$$\lambda_l = V_l + \sum_{l'} W_{ll'} n_{l'} + 2 \frac{\partial z_l^2}{\partial n_l} \bar{\varepsilon}_l - 2t \sum_{\gamma=\pm 1} z_{l+\gamma} \frac{\partial z_l}{\partial n_l} \bar{\chi}_l^{(\gamma)} - \mu, \qquad (4.31)$$

## 4.3 Slave-boson mean-field treatment

where we have introduced the following definitions

$$\bar{n}_l = \frac{2}{N_\parallel} \sum_{\mathbf{k}\nu} \psi_{\mathbf{k}\nu}(l)^2 f_T(E_{\mathbf{k}\nu}), \tag{4.32}$$

$$\bar{\varepsilon}_l = \frac{1}{N_\parallel} \sum_{\mathbf{k}\nu} \varepsilon_{\mathbf{k}} \psi_{\mathbf{k}\nu}(l)^2 f_T(E_{\mathbf{k}\nu}), \tag{4.33}$$

$$\bar{\chi}_l^{(\gamma)} = \frac{2}{N_\parallel} \sum_{\mathbf{k}\nu} \psi_{\mathbf{k}\nu}(l) \psi_{\mathbf{k}\nu}(l+\gamma) f_T(E_{\mathbf{k}\nu}). \tag{4.34}$$

The saddle-point equations (4.29)-(4.31) together with the effective Schrödinger equation (4.18) and the electrostatic potential (4.21) define the Schrödinger-Poisson-Gutzwiller problem which we solve numerically.

### 4.3.5 Numerical scheme

We use the algorithm schematically shown in Fig. 4.1 to solve the nonlinear saddle-point equations. It basically consists of first maximizing the free energy $f(n, d, \lambda)$ with respect to $\lambda$ and then minimizing the resulting function with respect to $d$ and $n$ (see, e.g., Ref. [114]). It turns out that this strategy leads to a stable algorithm in a wide range of parameters whereas an iterative solving of the saddle-point equations fails for larger values of $U$ due to the strong non-linearity. Thus, our algorithm contains basically two loops where in the first loop the free energy is maximized with respect to $\lambda$ and in the second loop the resulting function is minimized with respect to $d$ and $n$. The calculation of the free energy and of its gradients in the thermodynamic limit $N_\parallel \to \infty$ involves a two-dimensional $\mathbf{k}$-integration which can be reduced to a one-dimensional energy-integration including the density of states of the two-dimensional nearest neighbor tight-binding model

$$\rho_\sigma(\varepsilon) = \begin{cases} \frac{1}{2\pi^2 t} \mathrm{K}\left[1 - \left(\frac{\varepsilon}{4t}\right)^2\right] & \text{if } |\varepsilon| \leq 4t; \\ 0 & \text{else.} \end{cases} \tag{4.35}$$

Here, $\mathrm{K}(x)$ is the complete elliptic integral of the first kind. Note that there is a Van Hove singularity at $\varepsilon = 0$. For the diagonalization of Eq. (4.18) we impose open boundary conditions and take typically about 30 layers into account, depending on the number of ion layers $N$. All the calculations are performed at

$T = 0$. The fact that formally only the free energy dispersion $\varepsilon_\mathbf{k}$ enters into the SBA analysis means also that we can treat various lattice topologies in exactly the same framework with modified density of states.

## 4.4 Results and discussion

### 4.4.1 Ground-state properties

The dependence of the mean fields and of the hopping renormalization factor on the onsite interaction strength is summarized in Fig. 4.2 for the $N = 10$ heterostructure with $E_C = 0.8t$. The electronic charge distribution $n_l$ shown in panel (a) has a rather weak dependence on $U_r$: the effect of the long-range Coulomb field dominates over the short-range correlations [28, 39, 40]. Similarly, the Lagrange multiplier $\lambda_l$ shown in panel (c) varies only weakly with $U_r$. On the other hand, the fraction of doubly occupied sites [panel (b)] and the hopping renormalization [panel (d)] strongly depends on $U_r$.

**Double occupancy, hopping renormalization and local moments**

It is clear, that in order to reduce the onsite potential energy cost, doubly occupied sites are suppressed by increasing $U_r$. In the minimal interacting case, $U_r = 0$, the fraction of doubly occupied sites in layer $l$ is $d_l^2 = n_l^2/4$. With increasing $U_r$, $d_l$ tends to zero. Doubly occupied sites in the center of the heterostructure, where the charge is close to one, are strongest suppressed, see Fig. 4.2(b). Consequently, the in-plane-hopping renormalization factor $z_l^2$ shown in Fig. 4.2(d) is reduced there reflecting the suppression of charge fluctuations. On the other hand, in the (heavily) doped BI region, $d_l$ varies only little with increasing $U_r$ and stays close to the minimal interacting value. Therefore, $z_l^2 \approx 1$ in this region, irrespective of the value of $U_r$. The suppression of the double-occupancy is directly related to the formation of local moments. Indeed, from the Pauli principle it follows that

$$\delta m_l^2 \equiv \langle (n_{i\uparrow} - n_{i\downarrow})^2 \rangle = n_l - 2d_l^2, \quad i \in \text{layer}\, l. \tag{4.36}$$

However, in our simple treatment these local moments are absent. In other words, as we will critically discuss in Sec. 4.5, the paramagnetic SBA solution

4.4 Results and discussion

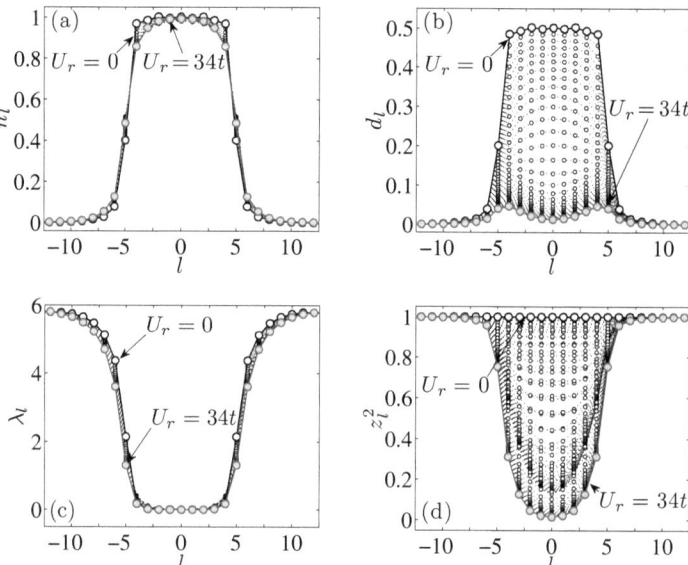

Figure 4.2: (a) The electronic charge distribution $n_l$, (b) the amplitude of the double occupancy $d_l$, (c) the Lagrange multiplier $\lambda_l$ and (d) the in-plane-hopping renormalization factor $z_l^2$ for the $N = 10$ ion-layers heterostructure with $E_C = 0.8t$. Different curves correspond to different values of the onsite repulsion $U_r = 0, \ldots, 34t$.

suggests a state where these local moments are screened by the Kondo effect.

**Electronic charge distribution**

In the paramagnetic phase discussed here, there are two competing trends to reduce the double occupancy and thus the onsite energy cost:

(i) Spreading of the electron charge distribution.

(ii) Localization of the electronic states.

Of course, in both cases, long-range Coulomb as well as kinetic energy have to be payed. Although we find a rather small dependence of $n_l$ on $U_r$, a closer look at the evolution of the charge distribution allows to distinguish two different regimes which are characterized by the different role of the two mechanisms, see Fig. 4.3(a). Starting from the minimal interacting case and increasing $U_r$ results in a spread of the particle density. This effect is also included in the Hartree approximation where the decoupling of the $U$-term leads to an additional term in the self-consistent electrostatic potential and $-e\phi_l \to -e\phi_l + U_r n_l/2$ which favors a homogeneous charge distribution. We therefore refer to this regime as the *Hartree regime*. But for all $N > 1$ we observe that in most layers the density $n_l$ evolves non-monotonically with increasing $U_r$. Namely, there is a turning point where the slope of $n_l(U_r)$ changes sign. Figure 4.3(a) shows the occupancy of the central layers $l = 0$, $l = \pm 1$ and $l = \pm 2$ of the $N = 10$ heterostructure as a function of $U_r$. The turning-point-value of $U_r$ is different for the different layers and varies also with $N$ but is roughly in the range of the critical value $U_c \approx 16t$ of the Brinkman-Rice transition in the homogeneous system. We therefore refer to the regime $U_r > U_c$ as the *Mott regime*.

**Penetration depth**

More information about the electronic charge distribution is gained by considering the following two quantities: We define a quantity measuring the width of the interface as

$$\Delta = \sum_l n_l(1 - n_l) \tag{4.37}$$

## 4.4 Results and discussion

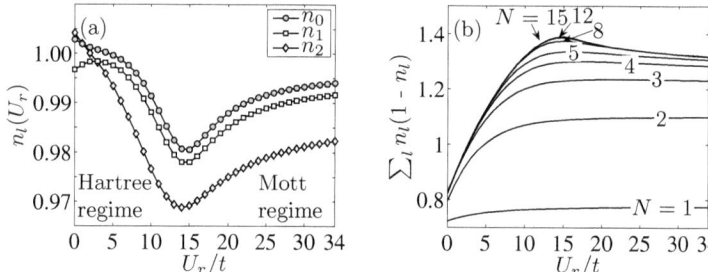

Figure 4.3: (a) The occupancy of the central layers as a function of $U_r$ for $N = 10$. The regime $U_r < U_c \approx 16t$ is called the "Hartree regime" while the regime $U_r > U_c$ is called the "Mott regime". (b) The quantity $\Delta = \sum_l n_l(1 - n_l)$ measuring the width of the interface as function of $U_r$ for different heterostructures $N$.

and the square mean distance of the charge from the center of the heterostructure

$$\bar{l}^2(N, U_r) = \frac{\sum_l l^2 n_l(N, U_r)}{N}. \tag{4.38}$$

$\Delta$ shown in Fig. 4.3(b) is a measure of the region where the electron density $n_l$ differs significantly from the bulk values $n_l = 0$ (BI) and $n_l = 1$ (MI). In the strongly interacting limit the mobile carriers are mainly confined to this region (see also Sec. 4.4.2). Figure 4.4 shows $\bar{l}^2$ for $N = 3, 5, 10$ as a function of $U_r$. The behavior of both quantities is different in the Hartree regime as compared to the Mott regime: in the Hartree regime we observe a rather steep increase of $\Delta$ and $\bar{l}^2$ indicating that the reduction of the doubly occupied sites is mainly achieved by the spreading of the charge, namely, by the mechanism (i). On the other hand, in the Mott regime, $\Delta$ and $\bar{l}^2$ is approximatively constant (it flattens or even slightly decreases) suggesting that in this regime the effect of localization [mechanism (ii)] compensates the mechanism (i). In addition, for $N > 5$, the width $\Delta$ of the interface becomes almost independent of $N$ and differs only in the vicinity of $U_c$ reflecting that for large $N$ the two interfaces are effectively disconnected.

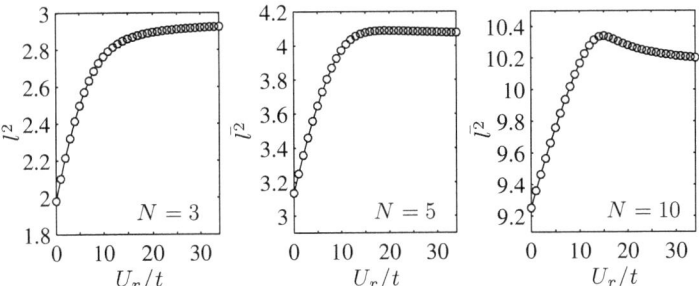

Figure 4.4: The square mean distance $\bar{l}^2$ of the electronic charge from the center as defined in Eq. (4.38) shown for $N = 3, 5, 10$ as a function of $U_r$.

The behavior of $\bar{l}^2$ and $\Delta$ is a consequence of the competition between the three energy scales in the problem, namely $E_C$, $U_r$ and the bandwidth $W = 12t$. In the Hartree regime $E_C < U_r \ll W$ the qualitative behavior observed in Figs. 4.3 and 4.4 is basically determined by the competition between $E_C$ and $U_r$. The penetration depth thus increases by increasing $U_r$ due to the mechanism (i). On the other hand, in the Mott regime $E_C < W \ll U_r$ the penetration depth is mainly determined by the competition between $E_C$ and $W$ and therefore essentially independent of $U_r$.

### 4.4.2 Metallic interfaces in the Mott regime

Within the present approach, quasi-two dimensional metallic behavior is expected in the weakly correlated Hartree regime. In the following we discuss how a strongly correlated metallic state emerges at the interface in the Mott regime by recapitulating (within the SBA approach) the arguments given by Okamoto and Millis [39] who based their reasoning on the DMFT picture.

## 4.4 Results and discussion

**Layer-resolved spectral weight**

In order to gain information about the spatial extend of the metallic phase we consider the layer-resolved single-particle spectral weight. In principle, the spectral functions are accessible in photoemission [110] or scanning tunneling microscopy.

The **k**-dependent layer resolved spectral density is defined as the imaginary part of a retarded one-particle Green function

$$A_{l\sigma}(\mathbf{k},\omega) = -\frac{1}{\pi}\mathrm{Im}\, G_{ll,\sigma}(\mathbf{k},\omega+i0^+). \tag{4.39}$$

The SBA approximation focuses on the coherent (low-energy) part of the spectral weight.[7] Nonetheless, the low-energy part is important for the metallicity of the heterostructure. Following Sec. 4.2 the *coherent* part of $G_{ll,\sigma}(\mathbf{k},\omega)$ is given by

$$G_{ll,\sigma}^{\mathrm{coh}}(\mathbf{k},\omega+i0^+) = \sum_\nu \frac{z_l^2 \psi_{\mathbf{k}\nu}(l)^2}{\omega - E_{\mathbf{k}\nu} + i0^+} \tag{4.40}$$

and we find for the coherent part of the spectral density

$$A_{l\sigma}^{\mathrm{coh}}(\mathbf{k},\omega) = \sum_\nu z_l^2 \psi_{\mathbf{k}\nu}(l)^2 \delta(\omega - E_{\mathbf{k}\nu}). \tag{4.41}$$

The ($\mathbf{k},\sigma$ integrated) layer resolved spectral density is given by

$$A_l^{\mathrm{coh}}(\omega) = \frac{1}{N_\parallel}\sum_{\mathbf{k}\sigma} A_{l\mathbf{k}\sigma}^{\mathrm{coh}}(\omega) = z_l^2 \rho_l^*(\omega), \tag{4.42}$$

where the local quasiparticle DOS $\rho_l^*(\omega)$ is given in (4.13). Figure 4.5 shows the layer resolved spectral density $A_l^{\mathrm{coh}}(\omega)$ in the minimal interacting case and for $U_r = 24t$ for the $N=10$ heterostructure. In the latter case, the coherent part is strongly reduced towards the center of the heterostructure. The missing spectral weight indicates nearly insulating behavior and that most of the spectral weight is transferred to upper and lower Hubbard band at a higher energy scale. However, outside the heterostructure, the spectral weight of the near-Fermi-surface-states

---

[7]In order to access to lower and upper Hubbard bands within slave-boson theory one has to go beyond the saddle-point approximation [99]. This issue is not touched here but is partially discussed in Sec. 3.3.

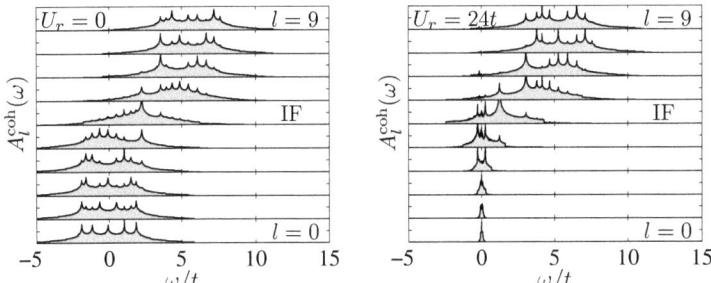

Figure 4.5: Coherent part of the layer resolved spectral density for $U_r = 0$ and $U_r = 24t$ for the $N = 10$ heterostructure showed for the layers $l = 0, \ldots, 9$. Here, $l = 0$ corresponds to the layer in the center of the heterostructure.

is only weakly suppressed and almost identical to the minimal interacting case. We find that in the doped BI region the chemical potential ($\omega = 0$) is, in all cases studied, pinned to the the bottom of the conduction band. The highest-lying electron states are therefore only weakly bound [40]. The various singularities showing up in the spectral density enter here because of the Van Hove singularity in the free electron density of states $\rho_\sigma(\varepsilon)$.

**Coherent charge density**

By integration of Eq. (4.42) we find the coherent part of the particle density

$$n_l^{\text{coh}} = \int_{-\infty}^{\infty} d\omega A_l^{\text{coh}}(\omega) f_0(\omega) = z_l^2 n_l. \qquad (4.43)$$

Figure 4.6 shows $n_l^{\text{coh}}$ for various values of the onsite interaction $U_r$. The coherent particle density clearly shows humps at the interfaces between Mott and band-insulating material whereas in the center of the heterostructure the coherent part is strongly suppressed for strong onsite interactions. These humps have a width of approximately $3a$ which suggests that the coherent quasiparticles responsible for the metallic behavior are confined to a very narrow region at the interface.

## 4.4 Results and discussion

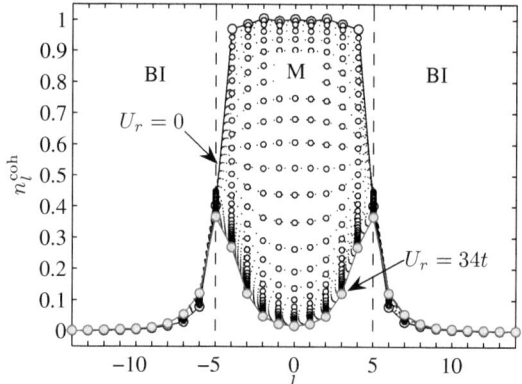

Figure 4.6: The coherent particle density $n_l^{\text{coh}}$ of the $N = 10$ heterostructure for various values of $U_r = 0, t, \ldots, 34t$.

For large $N$ and $U_r$ the coherent particle density is essentially independent of $N$ and $U_r$.

### 4.4.3 Quasiparticle properties

In this section we will discuss the properties of the coherent quasiparticles which in the Mott regime are located at the interfaces. Whenever it is appropriate, we will use the notation $E_\nu(\varepsilon)$ for the quasiparticle dispersion and $\psi_{\varepsilon\nu}(l)$ for the envelope wave function because the momentum dependence enters only through $\varepsilon \equiv \varepsilon_\mathbf{k} = -2t(\cos k_x + \cos k_y)$.

**Quasiparticle weight and effective mass**

Integration of $A_{l\sigma}(\mathbf{k}, \omega) f_0(\omega)$ over $l, \omega$ yields the ground-state momentum distribution function of the correlated electrons. The coherent part of the momentum distribution function $n_{\mathbf{k}\sigma}^{\text{coh}} = \sum_\nu n_{\mathbf{k}\nu\sigma}^{\text{coh}}$ contains contributions from the different

subbands
$$n^{\text{coh}}_{\mathbf{k}\nu\sigma} \equiv n^{\text{coh}}_{\mathbf{k}\nu} = Z_{\mathbf{k}\nu} f_0(E_{\mathbf{k}\nu}). \qquad (4.44)$$

From Eq. (4.44) it follows that each state $(\mathbf{k}\nu\sigma)$ is weighted by a factor $Z_{\mathbf{k}\nu} \leq 1$ where

$$Z_{\mathbf{k}\nu} = \frac{\partial E_{\mathbf{k}\nu}}{\partial \varepsilon_{\mathbf{k}}} = \sum_l z_l^2 \psi_{\mathbf{k}\nu}(l)^2. \qquad (4.45)$$

Therefore, also the discontinuity of $n^{\text{coh}}_{\mathbf{k}\nu\sigma}$ at the Fermi energy is reduced, namely, it is equal to

$$Z_\nu = \frac{\partial E_{\mathbf{k}\nu}}{\partial \varepsilon_{\mathbf{k}}}\bigg|_{E_{\mathbf{k}\nu}=0} = \sum_l z_l^2 \psi_{\varepsilon^*_\nu \nu}(l)^2. \qquad (4.46)$$

Here, $\varepsilon^*_\nu$ is defined by the equation

$$E_\nu(\varepsilon^*_\nu) = 0. \qquad (4.47)$$

The quasiparticle weight $Z_\nu$ is a characteristic quantity describing correlation effects. In the local Fermi liquid picture it is equal to the mass enhancement factor showing up as a reduction of the Fermi velocity of the sheet $\nu$. Consequently, it also enters the linear specific heat at low temperatures by the enhancement of the quasiparticle DOS $\rho^*(0)$ [see Eq. (4.14)]

$$c_v = -T\frac{\partial^2 f}{\partial T^2} = \frac{\pi^2 k_B^2}{3}\rho^*(0)T$$

where $\rho^*(0) = \sum_l \rho_l^*(0)$ is the total quasiparticle DOS at the Fermi energy.

The quasiparticle weights $Z_\nu$ of the partially filled subbands of the $N = 3$ heterostructure are shown in Fig. 4.7. Energetically lower lying subbands are in general more affected by $U_r$ and the $\nu = 1$ subband always has the smallest quasiparticle weight. Note, however, that this subband is *not* closest to half filling. The exact order of $Z_\nu$ for $\nu > 1$ depends on $N$ and in some cases also on $U_r$. Because odd wave functions have nodes in the center of the quantum well at $z = 0$, a parity effect is visible: If $N$ is even the electronic sites are at $l = 0, \pm 1, \pm 2, \ldots$ and odd-parity subbands are less renormalized because $\psi_{\varepsilon^*_\nu \nu}(0) = 0$. On the other hand, if $N$ is odd, the electronic sites are at $l = \pm 1/2, \pm 3/2 \ldots$ and the odd-parity subbands tend to be more strongly renormalized because the zero at $z = 0$ lies between the electronic sites. For $N = 3$, this parity effect

## 4.4 Results and discussion

is seen in Fig. 4.7(b) where $Z_4 < Z_3$ and $Z_6 < Z_5$. Also shown in Fig. 4.7(b) is a comparison between the quasiparticle weights obtained in the SBA and the values of the two-site DMFT calculations performed by Okamoto and Millis [39]. The qualitative behavior is very similar. Nevertheless, it appears that the SBA systematically yields higher values for the quasiparticle weights. We will come back to this point in Sec. 4.6.1.

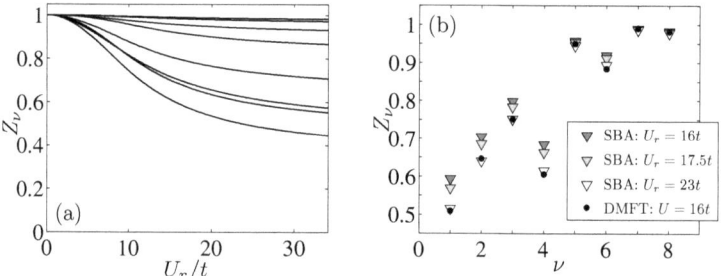

Figure 4.7: (a) Quasiparticle weight $Z_\nu$ of the partially filled subbands $\nu = 1, \ldots, 8$ as a function of $U_r$ for $N = 3$. (b) Comparison of the quasiparticle weight in the $N = 3$ heterostructure obtained within the two-site dynamical mean-field theory (DMFT) (courteously taken from Ref. [39]) and the Kotliar and Ruckenstein slave-boson mean-field approximation (SBA) for fixed values of the interaction strength.

**Approaching bulk**

Increasing the number $N$ of counter-ion layers has a profound effect on the strength of the renormalization. While it is only modest for $N = 3$, it can become substantial for certain subbands when introducing more counter-ion layers. We show in the right panel of Fig. 4.8 the quasiparticle weight $Z_1$ of the lowest lying subband as a function of $U_r$ for various values of $N$. For $N = 1$ there is no significant renormalization of the quasiparticle weight which reflects the special role of this case [40]. To have a more quantitative description how onsite correlation effects increase with growing $N$ we define a characteristic value

$\tilde{U}(N)$ for each $N$ by extrapolating $Z_1(U_r)$ at the inflection point to zero which yields $\tilde{U}(N)$. In the left panel of Fig. 4.8 we show $\tilde{U}(N)$ as a function of $N$. For $N \to \infty$ we find that $\tilde{U}(N)$ saturates at $U_c \approx 16t$, the critical interaction strength for the Mott transition in the bulk system of the sandwiched material.

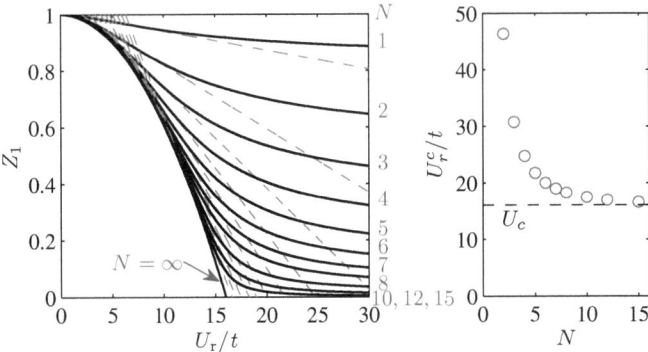

Figure 4.8: Left: The quasiparticle weight $Z_1$ for $N = 1, 2, \ldots, 8, 10, 12, 15, \infty$ as a function of $U_r$ (solid curves). In order to define a characteristic value $\tilde{U}(N)$ we extrapolate $Z_1$ at the inflection point to zero (dashed lines). Right: $\tilde{U}(N)$ as a function of $N$. The dashed line denotes the value of the Mott transition in the $N \to \infty$ model.

**Quasiparticle dispersion**

For $E_C = 0.8t$ we find in general $N + 5$ partially filled subbands in the ground state. In the minimal interacting limit ($U_r = 0$), the dispersion of these subbands has the form $E_{\mathbf{k}\nu} = E_\nu + \varepsilon_\mathbf{k}$, i.e. $E_\nu(\varepsilon_\mathbf{k})$ has slope one. Increasing $U_r$ leads to a renormalization of the quasiparticle dispersions and $E_\nu(\varepsilon)$ flattens, thus indicating localization of the electronic states. From Eq. (4.45) it can be deduced that the slope of $E_\nu(\varepsilon_\mathbf{k})$, namely $Z_{\mathbf{k}\nu}$, is in general less than one.

As an example we consider the $N = 3$ heterostructure with 8 partially filled subbands and the $N = 8$ heterostructure with 13 partially filled subbands. The

## 4.4 Results and discussion

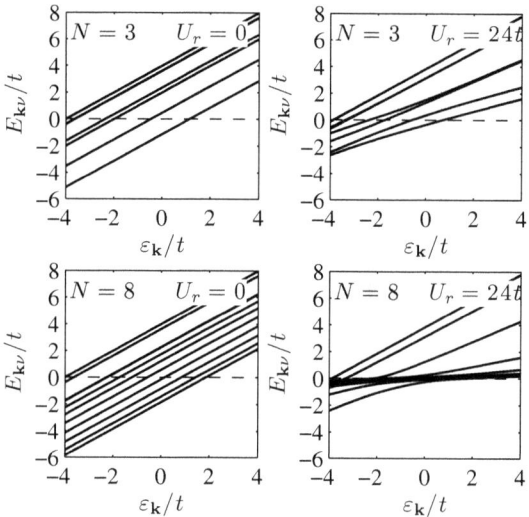

Figure 4.9: The quasiparticle dispersion $E_{\mathbf{k}\nu}$ of the partially filled subbands in the $N = 3$ and $N = 8$ heterostructure as a function of $\varepsilon_{\mathbf{k}} = -2t(\cos k_x + \cos k_y)$ both for $U_r = 0$ and $U_r = 24t$.

dispersion in the minimal interacting case and for $U_r = 24t$ is plotted in Fig. 4.9. The particular geometry of the system, namely the symmetry $z \leftrightarrow -z$, allows that some of the quasiparticle subbands are (almost) degenerate. We can distinguish two types of degeneracies: (1) Essentially independent of the interaction strength we find that the almost unbounded states (the highest partially filled subbands) give rise to pairwise degenerate subbands. In the case $N = 3$ this is observed for the states $\nu = 5, 6$ and $\nu = 7, 8$. A similar feature can be seen for the $N = 8$ heterostructure where the six highest subbands $\nu = 8 - 13$ are pairwise degenerate. (2) In addition, by increasing $U_r$, also the lowest subbands can become pairwise degenerate in certain $\mathbf{k}$-regions. This is observed in Fig. 4.9 for the $\nu = 1, 2$ and $\nu = 3, 4$ subbands of the $N = 8$ heterostructure and reflects that the two interfaces become disconnected from each other.

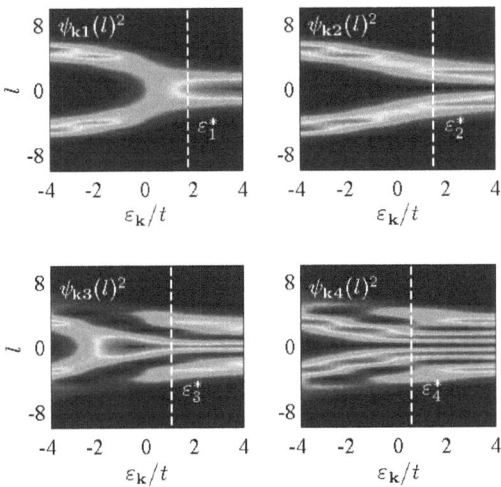

Figure 4.10: Density plot of the square of the four lowest lying transverse wave functions $\psi_{\mathbf{k}\nu}(l)^2$ as a function of $l$ and $\varepsilon_{\mathbf{k}}$ for $N = 10$ and $U_r = 14t$. The Fermi energy in the $\varepsilon_{\mathbf{k}}$-space is indicated as white dashed lines.

A better understanding for the quasiparticle states associated with the different subbands is obtained by considering the probability distribution $\psi_{\mathbf{k}\nu}(l)^2$ for different values of $\varepsilon_{\mathbf{k}}$. The rather complex dependence of the transverse part $\psi_{\varepsilon\nu}$ on $\varepsilon$ is shown in Fig. 4.10 by a density plot of $\psi_{\varepsilon\nu}(l)^2$ for $\nu = 1, \ldots, 4$ as function of $\varepsilon$ and $l$. In this plot, we interpolate $\psi_{\varepsilon\nu}(l)^2$ between different $l$'s. The $\varepsilon$ value of the Fermi energy is denoted by $\varepsilon_\nu^*$ and is indicated as a white dotted line. As seen in these color plots the character of the wave function changes dramatically when changing $\varepsilon_{\mathbf{k}}$ from $-4t$ to $4t$. Within the SBA this behavior can be understood as a result of the term $(z_l^2 - 1)\varepsilon_{\mathbf{k}}$ in the effective Schrödinger equation (4.18) which we can combine to a $\varepsilon_{\mathbf{k}}$-dependent effective potential $\lambda_l + (z_l^2 - 1)\varepsilon_{\mathbf{k}}$. This potential includes the screened ion-potential as

4.5 Interfacial heavy-fermion scenario

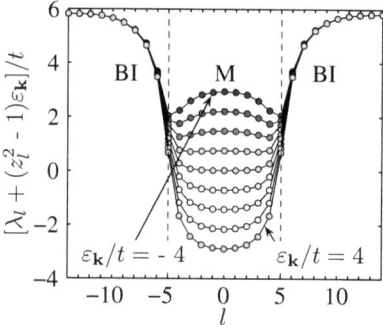

Figure 4.11: The effective one-dimensional potential $\lambda_l + (z_l^2 - 1)\varepsilon_{\mathbf{k}}$ for various values of $\varepsilon_{\mathbf{k}} = -4t, -3t, \ldots, 4t$ for $U_r = 14t$ and $N = 10$.

well as a "Mott potential" introducing an effective central barrier for $\varepsilon < 0$. As shown in Fig. 4.11, a double-well potential is formed for $\varepsilon_{\mathbf{k}} < 0$ and the regions $l > 0$ and $l < 0$ are almost decoupled. This means that electrons which have a low (in-plane) kinetic energy prefer to stay mainly in the interface region rather than in the center where the effect of the onsite interaction $U_r$ is much stronger.

**Subband filling**

Eventually, we remark here that the filling of the quasiparticle subbands also changes by varying $U_r$. However, this effect is rather weak and we refrain from discussing this point in more details.

## 4.5 Interfacial heavy-fermion scenario

We point out that we do not observe a strict Mott transition for finite $N$ and $U_r$ in the sense of a vanishing local spectral weight at the Fermi energy for at least one layer. Instead, the mutual doping of Mott and band insulator leads to metallic behavior of the whole structure. The system gains energy by keeping a certain amount of charge fluctuation meaning that $Z_\nu > 0$. In other words, the hopping between the individual layers is never renormalized to 0 and $n_l < 1$ and $z_l > 0$. In the spatially nonuniform system discussed here, there is a large amount of freedom to reduce the onsite energy cost. Among all the states characterized by $(\mathbf{k}\nu)$ only a part is strongly affected by the correlations. This allows for an optimal reduction of the onsite energy cost without cutting back much from the kinetic or long range Coulomb energy. Thus, the possibility that correlation effects can be selectively split among a large number of different states prevents the system from undergoing a strict Mott transition.[8]

Although our calculations for the paramagnetic phase reveal an overall metallic state, the appearance of almost flat subbands, see Sec. 4.4.3, clearly indicates the emergence of local moments in the central region of the heterostructure. Interlayer hopping introduces a coupling (hybridization) between these almost localized degrees of freedom and the more itinerant ones at the interface which leads to an overall metallic state with a substantial renormalization occurring for the quasiparticle states having wave functions predominantly located in the central region of the heterostructure. This behavior resembles in certain aspects that of the strongly renormalized quasiparticle states encountered in heavy Fermion systems, as described by the periodic Anderson model (PAM), in which the electronic properties are also governed by strongly correlated and localized orbitals hybridizing with orbitals of a wide conduction band [115–117]. In analogy, we shall use the terminology *interfacial heavy-fermion* state to describe the local Fermi liquid obtained in the SBA approximation of the heterostructure model. Roughly speaking, the interfacial heavy-fermion scenario consists of two steps: First, electronic charge transfer to the interface gives rise to itinerant electronic states. Second, these itinerant states weakly mix with the (almost) localized degrees of freedom in the adjacent layers resulting in a heavy Fermi liquid. It is possible that the Kondo effect [118] might play a role in screening the nearby local moments and to stabilize the interfacial heavy-fermion state. Nevertheless, understanding the electronic properties at the interface clearly poses a new challenge without analogy to intermetallic heavy-fermion compounds associated with the presence of localized $f$-orbitals. For example, in the present situation, the

---

[8]Nevertheless, introducing a gate voltage to tune the chemical potential $\mu$ one expects that a Mott insulating state in the center of the heterostructure can be stabilized also in the paramagnetic phase.

same orbitals give rise to localized and itinerant degrees of freedom, depending on the occupation, and there is a direct, antiferromagnetic exchange interaction between the localized moments. Furthermore, it is likely that due to the reduced dimensionality spin fluctuations are enhanced and may play a dominant role. In any case, from these considerations it becomes clear that the treatment of the local moments with possibly short-ranged magnetic correlations has to be improved compared to its rather poor description in the SBA.

## 4.6 Comparison to related approaches

### 4.6.1 Comparison to two-site DMFT

In this section we compare our results more closely to the ones reported by Okamoto and Millis in Ref. [39] who used the two-site approximation of the dynamical mean-field theory [120] to study the same model. It is interesting to note that within the two-site DMFT the quasiparticle weight of the homogeneous system at half filling, $n = 1$, as function of $U/U_c \leq 1$ is equal to the SBA result

$$Z = 1 - \left(\frac{U}{U_c}\right)^2. \qquad (4.48)$$

What differs is the expression for the critical interaction strength for the Mott transition. In the SBA approximation one recovers the Brinkman-Rice criterion (3.23) where the *first* moment of the free DOS times the distribution function determines $U_c$ In the two-site DMFT it is the *second* moment of the density of states which enters [120]. Typically, $U_c$ is about 10% larger than the critical point for the Mott transition obtained from the two-site DMFT, depending on the free electron density of states [121]. The equivalence of $Z(U/U_c)$ breaks down if $n \neq 1$. By comparing the two-site DMFT results of Ref. [120] for $Z$ away from half filling with the Gutzwiller result obtained by the SBA scheme we find the general trend that for systems away from half filling the discrepancy between the two methods increase with increasing $U$. For $U$ well above $U_c$ the SBA gives up to $40 - 50\%$ larger quasiparticle weights as compared to the two-site DMFT.

It is therefore an interesting question to which extend the two methods agree when applied to a spatially non-uniform system where the charge density varies locally between $n_l \approx 0$ and $n_l \approx 1$. In this context, the distinction between $U_r$ and $U$ does not play any significant role. In Fig. 4.7(b) we compare the quasiparticle weights of the $N = 3$ heterostructure calculated within the two-site DMFT for $U = 16t$ with the SBA results for $U_r = 16t, 17.5t, 23t$. Best agreement between the two methods is achieved for $U_r = 23t \approx 1.45U$.

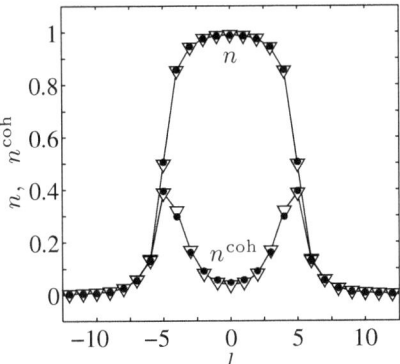

Figure 4.12: The total and the coherent part of the charge profile of the $N = 10$ heterostructure calculated within the two-site DMFT (Ref. [39]) for $U = 16t$ (circles) and within the SBA approximation for $U_r = 23t$ (triangles).

A similar observation is also made when comparing the results of the $N = 10$ heterostructure. In Fig. 4.12 we show the total and the coherent part of the charge profile of the $N = 10$ heterostructure calculated within the two-site DMFT for $U = 16t$ and the SBA approximation for $U_r = 23t$. Whereas the total charge distribution remains nearly unchanged if $U_r$ exceeds $16t$ and agrees within a few percent with the two-site DMFT results it is more subtle to compare the coherent charge distribution $n_l^{\text{coh}}$ [Eq. (4.43)]. We find best agreement for $U_r = 23t$. This value is roughly 45% larger than the value of $U = 16t$ used in the two-site DMFT calculation and, interestingly, coincides with the one found for the $N = 3$ system by comparing the quasiparticle weights [Fig. 4.7(b)].

## 4.6 Comparison to related approaches

In both cases contrasted in this section we find that the SBA gives larger quasiparticle weights for the same onsite interaction strength as compared to the two-site DMFT. The qualitative effect of $U_r$, however, is in both methods the same. Furthermore, for the considered cases the results of the SBA for a value of $U_r$ sufficiently larger than $U$ agree within a few percent with the two-site DMFT results for $U$. This observations are consistent with the increasing discrepancy between the two methods observed in the homogeneous system away from half filling.

### 4.6.2 Comparison to the "pseudocanonical" GA

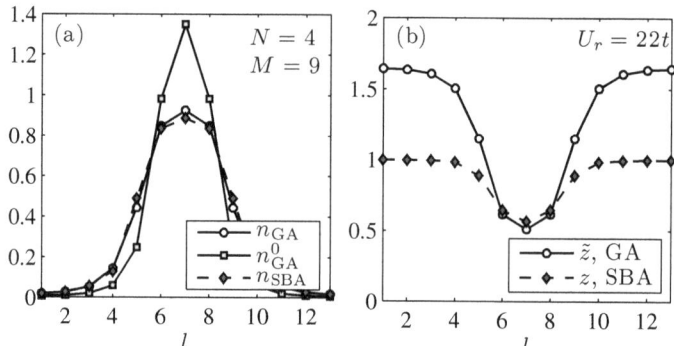

Figure 4.13: Comparison between the "pseudocanonical" Gutzwiller approximation (GA) and the Kotliar and Ruckenstein slave-boson meanfield approximation (SBA) for a Hubbard superlattice with period $N + M = 13$ and $N = 4$ counter-ion layers. (a) The projected charge density $n_l$ and the unprojected charge density $n_l^0$ in the GA as well as the charge density in the SBA. (b) The hopping renormalization factor obtained by the two approaches.

As already mentioned in Sec. 3.2.1 there is a close relation between the SBA and the Gutzwiller approximation (GA) scheme. Indeed, for many homogeneous systems, the equivalence between the GA and the SBA has been proven, see e.g. [122]. In spatially non-uniform systems, however, GA and SBA yield different results because the *projected charge* distribution $n_l$ is in general *not equal* to the *unprojected distribution* $n_l^0$ [123]. By introducing local fugacities it is possible to constrain the *local charge* such that $n_l^0 = n_l$ and one recovers the SBA equations. However, the locally constrained approach yields higher energies than the "pseudocanonical" scheme where only the *total charge* of the projected and the unprojected state is fixed. It is therefore of interest to discuss differences between the SBA and this pseudocanonical GA and, indeed, for the Hubbard heterostructure we find some interesting differences, see Fig. 4.13 and 4.14. The energy functional in the GA is essentially equal to (4.27) with the important difference that in the effective Hamiltonian (4.18) the hopping renormalization factor $z_l$ is replaced by

$$\tilde{z}_l(n_l, d_l, n_l^0) = \frac{\sqrt{(1 - n_l + d_l^2)(n_l - 2d_l^2)} + d_l\sqrt{n_l - 2d_l^2}}{\sqrt{n_l^0(1 - n_l^0/2)}}. \quad (4.49)$$

Note that $\tilde{z}_l$ is in general *not restricted* to values less or equal to one. Furthermore, a global Lagrange multiplier $\alpha$ is introduced such that the total projected charge is equal to the total unprojected charge

$$\sum_l n_l = \sum_l n_l^0 \equiv N. \quad (4.50)$$

It is then possible to completely eliminate $n_l^0$ from the Gutzwiller energy functional by the relation [124]

$$n_l^0 = 2\frac{e^{-\alpha/2}(n_l/2 - d_l^2)}{e^{\alpha/2}(1 - n_l + d_l^2) + e^{-\alpha/2}(n_l/2 - d_l^2)} \quad (4.51)$$

where $\alpha$ satisfies

$$2\sum_l \frac{e^{-\alpha/2}(n_l/2 - d_l^2)}{e^{\alpha/2}(1 - n_l + d_l^2) + e^{-\alpha/2}(n_l/2 - d_l^2)} = N. \quad (4.52)$$

In Fig. 4.13 we compare the charge density and the hopping renormalization factor obtained within the two schemes for a Hubbard superlattice with period $L = N + M = 13$ and $N = 4$ counter-ion layers. While the overall charge density

## 4.6 Comparison to related approaches

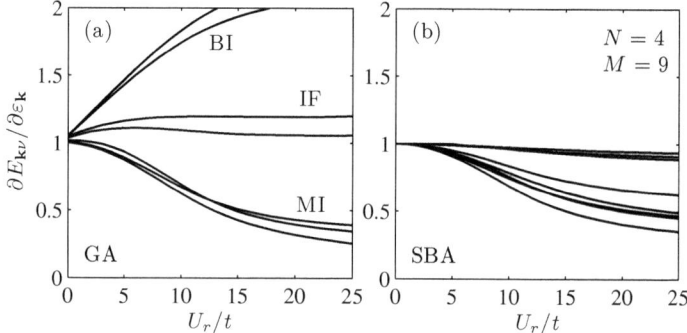

Figure 4.14: Comparison of $\partial E_{\mathbf{k}\nu}/\partial\varepsilon_{\mathbf{k}}$ for the partially filled subbands obtained in (a) the Gutzwiller approximation (GA) and (b) the Kotliar and Ruckenstein slave-boson mean-field approximation (SBA) for a Hubbard superlattice with period $N + M = 13$ and $N = 4$ counter-ion layers. Note that in the SBA, $\partial E_{\mathbf{k}\nu}/\partial\varepsilon_{\mathbf{k}}$ can be interpreted as the inverse effective mass enhancement and as quasiparticle weight. This straightforeward interpretation fails in the GA.

is similar in both approaches, the layer-resolved hopping renormalization factor shows striking differences. Clearly, $\tilde{z}_l > 1$ in the BI region, and the straightforeward interpretation of $\tilde{z}_l^2$ as a local quasiparticle weight (which is valid in the SBA) is no-longer true in the GA. As a consequence of the spatial dependence of $\tilde{z}_l$ we find that the derivative of the one-particle dispersion $\partial E_{\mathbf{k}\nu}/\partial\varepsilon_{\mathbf{k}} > 1$ for the states located in the BI and at the interface region (IF), see Fig. 4.14. At the present stage, the physical content of these observations is not yet clear and it is desirable to study the properties of the quasiparticle excitations (effective mass, spectral weight etc.) in the GA in more details.

## 4.7 Conclusions

To summarize, we have presented a mean-field study of a model heterostructure characterized by a BI/M/BI stacking (see Fig. 2.4) based on the Kotliar and Ruckenstein slave-boson mean-field approximation. Although the model discussed here is too simple to make quantitative statements about the experimentally realized $LaTiO_3/SrTiO_3$ interface it allows to discuss some aspects of short-range correlation in spatially nonuniform systems. In particular, it suggests that the mobile carriers are confined to a relatively narrow region at the interface [39] forming a quasi-two-dimensional electron gas as experimentally observed in $LaTiO_3/SrTiO_3$ heterostructures [13].

The SBA offers a relatively simple and transparent way to include both the self-consistent determination of the charge distribution as well as spatially varying hopping renormalization factors due to strong correlations. The basis of this approach is an effective one-dimensional Schrödinger equation in combination with self-consistency equations for the charge distribution and the double occupancy. Due to the lack of spatial homogeneity the quasiparticles obtain a nontrivial momentum dependence which can vary substantially by changing the onsite interaction $U$ or the width $N$ of the sandwiched material. This dependence was studied by calculating the renormalization of the quasiparticle dispersion and the quasiparticle weight and was interpreted as the fingerprints of an interfacial heavy Fermi state. To which extend this scenario is applicable in real systems requires further studies. Whereas $Sr_{1-x}La_xTiO_3$ shows clear Fermi liquid behavior with enhanced electron-electron scattering processes and a Wilson ratio of approximately 2 as $x \to 1$ (Ref. [125]) the experimental data available for $LaTiO_3/SrTiO_3$ superlattices and heterostructures are less systematic but are at least consistent with Fermi liquid behavior [111, 126]. From the theoretical point of view, it is an open problem if an interfacial Kondo effect can be realized in certain situations when treating the magnetic degrees of freedom more accurately. However, in the SBA analysis of the model, the system *is* an interfacial heavy Fermi liquid with well defined quasiparticles and the present approach offers a rout to access to Fermi liquid parameters on the basis of a microscopic model. On the other hand, we confirmed various results reported in previous DMFT studies [39] which leads to the conclusion that the SBA gives a reasonable description

## 4.7 Conclusions

of the low-energy part and is thus able to catch the most important aspects of correlations in spatially nonuniform situations. However, only onsite correlations can be taken into account and the introduction of a layer dependent mass renormalization distinctly increases the computational effort as compared to Hartree mean-field calculations. But for all that, compared to DMFT calculations, the SBA approximation has the advantage that it is numerically less involved.

So far we did not treat other important aspects inherent in strongly correlated electron systems. The most conspicuous is the ignorance of magnetic correlations which especially in bipartite lattices is of relevance. To some extent this can be included also on the mean-field level by considering an enlarged unit cell. On the other hand, the results shown in this chapter are basically independent of the details of the band structure and may be suited to cases where magnetic order is suppressed - like on a triangular lattice. Nevertheless, inter-site correlations, namely spin-spin interaction between the emergent local moments, cannot be treated within the present approach. This is of particular interest in connection with superconductivity arising in doped Mott insulators.

# Chapter 5

# Optical conductivity and thermoelectricity in correlated superlattices

> We investigate electrical and thermal transport properties in atomically precise metallic superlattices based on strongly-correlated electron systems. The main focus is put on the free carrier optical response and on the Seebeck coefficient. At low temperatures, we study the role of multiple subband renormalization in the quasiparticle transport using the semiclassical Boltzmann theory in the relaxation-time approximation. We find that features of the electronic structure reminiscent of heavy fermion physics provide favorable conditions to enhance the low-temperature thermopower. At high temperatures a generalization of Heikes formula for strongly correlated superlattices is used to estimate the entropic contribution to the Seebeck coefficient.

## 5.1 Introduction

In this chapter we are exclusively concerned with the experimentally observed metallic properties of the interface between Mott insulator (MI) and band insulator (BI) as for example realized in the $SrTiO_3/LaTiO_3$ system [13, 14, 111, 126].

## 5.1 Introduction

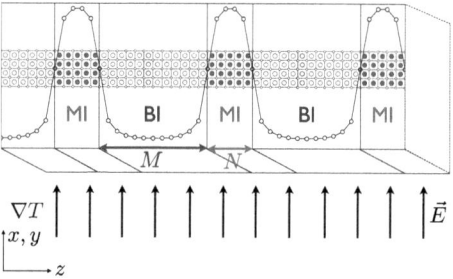

Figure 5.1: Schematic view of the setup considered in this chapter. $N$ is the number of unit cells of the Mott insulating material (MI) and $M$ is the number of unit cells of the band insulating material (BI). In addition, the conduction-band density is indicated. We assume a thermal gradient $\nabla T$ or an electrochemical field $\vec{E}$ along the $x$ direction.

We attempt to describe the transport properties of these quasi-two-dimensional metallic electron systems on a qualitative level and to clarify the role of the expected strong electron-electron interaction. We show that experimental data [111] on the optical conductivity of LTO/ STO superlattices are consistent within a simple theory for reasonable parameters. In addition, we focus on the low-temperature thermoelectrical effect. The motivation to study the thermoelectricity is based on several interesting observations: (i) The solid solution $Sr_{1-x}La_xTiO_3$ shows a large thermoelectric response [127] for $0 \leq x \leq 0.1$. (ii) From the study of narrow-band organic conductors [128], intermetallic compounds [129–131], as well as cobalt or other transition-metal oxides [132], it is known that correlation effects can enhance thermoelectricity. (iii) Lower-dimensional structures, such as quantum-well superlattices, offer additional parameters to optimize the thermoelectric response and, at best, overtop the bulk properties [133, 134].

A schematic view of the considered setup is shown in Fig. 5.1. We assume a cubic lattice structure with perfect lattice match. The artificial nanoscale structures with modulation in $z$ direction are composed of Mott insulating (MI) and band insulating (BI) bulk materials in a superlattice geometry with period $L = N + M$ and $N$ positively charged counter-ion layers. Charge neutrality requires that there are $N$ electrons per super unit cell. This setup is motivated by experiments on $(LTO)_N/(STO)_M$ superlattices but other material combinations may be possible. Quite in general we assume a polar perovskite Mott insulator A'BO$_3$ with a $3d^1$ configuration and a nonpolar perovskite band insulator ABO$_3$ with a $3d^0$ configuration. We are interested in transport properties perpendicular to the super cell modulation: electrical and thermoelectrical conductivities were calculated in response to an electrochemical field **E** or a temperature gradient $\nabla T$ in the $x$ direction. This is in contrast to previous theoretical work where the transport properties in strongly-correlated multilayer nanostructures along the $z$ direction was studied [43].

## 5.2 Drude weight and optical conductivity

The low frequency spectral weight of the (dissipative) optical conductivity $\sigma'(\omega)$ is one of the most relevant quantities both experimentally and theoretically to distinguish between an insulator and a metal [11, 135]. In a metal, a peak in $\sigma'(\omega)$ around $\omega = 0$ is observed whereas in insulators this so-called Drude peak is absent and a gap in the conductivity spectrum is found. The spectral weight of the Drude peak - the Drude weight $D$ - provides an order parameter for the metal-insulator transition. The system is metallic if $D > 0$ and insulating if $D = 0$. Studying the Drude weight therefore allows for an assessment of the metallicity in correlated nanoscale structures. The optical conductivity has recently been measured in STO/LTO superlattices [111] and below we compare our calculations to these measurements.

For a system with a (discrete) translation invariance (absence of disorder) in e.g. $x$-direction the real part of the complex optical conductivity $\sigma = \sigma' + i\sigma''$ at $T = 0$ has the form [135, 136]

$$\sigma'(\omega) = D\delta(\omega) + \sigma'_{\text{reg}}(\omega) \qquad (5.1)$$

with $\lim_{\omega \to 0} \omega \sigma'_{\text{reg}}(\omega) = 0$. The weight of the Dirac delta function $\delta(\omega)$ is identified as the Drude weight $D$. Before we discuss the quasiparticle contributions to $\sigma'(\omega)$ in correlated superlattices we briefly recall the general linear response formalism for the optical conductivity as well as Kohn's interpretation of the Drude weight as a stiffness constant with respect to twisted boundary conditions.

## 5.2 Drude weight and optical conductivity

### 5.2.1 Linear response

In the following we choose units such that $\hbar = 1 = c$ and we denote the charge of an electron as $-e$. The optical conductivity of the system is obtained in linear response to a time dependent uniform electric field in the $x$-direction as [137]

$$\sigma(\omega) = \frac{i}{\omega + i0^+} \left[ \frac{e^2}{N_s} \langle -\hat{T}_x \rangle_0 + \chi(\omega) \right] \quad (5.2)$$

where the kinetic energy operator in $x$ direction is given by

$$\hat{T}_x = -t \sum_{i\sigma} \left( c_{i\sigma}^\dagger c_{i+\mathbf{e}_x \sigma} + \text{h.c.} \right) \quad (5.3)$$

and $\langle \ldots \rangle_0$ denotes the ground-state expectation value. The current-current correlation function $\chi(\omega)$ has the following spectral representation in terms of eigenstates $|n\rangle$ of the system

$$\chi(\omega) = \frac{1}{N_s} \sum_n \left| \langle n | \hat{J}_{px} | 0 \rangle \right|^2 \left[ \frac{1}{\omega - \omega_{n0} + i0^+} - \frac{1}{\omega + \omega_{n0} + i0^+} \right] \quad (5.4)$$

with $\omega_{n0} = E_n - E_0$. The paramagnetic current in $x$ direction is given by

$$\hat{J}_{px} = iet \sum_{i\sigma} \left( c_{i\sigma}^\dagger c_{i-\mathbf{e}_x \sigma} - c_{i\sigma}^\dagger c_{i+\mathbf{e}_x \sigma} \right). \quad (5.5)$$

We decompose $\chi$ into real and imaginary part, $\chi = \chi' + i\chi''$. The real part of $\sigma(\omega)$ is found to be of the form of Eq. (5.1) with the Drude weight given by

$$D = \frac{\pi e^2}{N_s} \langle -\hat{T}_x \rangle_0 + \pi \chi'(0) \quad (5.6)$$

and a regular part $\sigma'_{\text{reg}}(\omega) = -\chi''(\omega)/\omega$. The real part of the optical conductivity obeys the well-known $f$-sum rule

$$\int_0^\infty \sigma'(\omega) d\omega = \frac{\pi e^2}{2N_s} \langle -\hat{T}_x \rangle_0 \quad (5.7)$$

implying in addition that

$$-\frac{\pi \chi'(0)}{2} = \int_0^\infty \sigma'_{\text{reg}}(\omega) d\omega.$$

### 5.2.2 Twisted boundary conditions

An alternative way to obtain the Drude weight was pointed out by Kohn [135]. The idea is to consider periodic boundary conditions in e.g. $x$-direction, thread the system with a flux $\Phi$ and study the dependence of the ground-state energy density $E_G(\Phi)$ on $\Phi$. The stiffness with respect to the twist is then proportional to the Drude weight $D$. Explicitly, we represent the flux by a vector potential

$$\mathbf{A} = (1/e)\Phi\hat{\mathbf{x}}/N_x \tag{5.8}$$

where $N_x$ is the number of lattice sites in $x$-direction. Then, due to the Peierls phase, the hopping in $x$-direction obtains an additional phase factor $\exp(\pm i\Phi/N_x)$ which modifies the free dispersion

$$\varepsilon_{\mathbf{k}}(\Phi) = -2t\left[\cos(k_x + \Phi/N_x) + \cos(k_y)\right]. \tag{5.9}$$

The stiffness constant is obtained in second order perturbation theory in $\Phi$ and comparison with (5.6) shows that the Drude weight can be calculated from the ground-state energy density $E_G(\Phi)$ as [135, 136, 138]

$$D = \pi e^2 N_x^2 \left.\frac{d^2 E_G(\Phi)}{d\Phi^2}\right|_{\Phi=0}. \tag{5.10}$$

We note here that the limit $\Phi \to 0$ has to be taken before going to the thermodynamic limit $N_x \to \infty$.

### 5.2.3 Quasi-particle contribution

The optical conductivity of strongly correlated materials usually shows a broad feature around a frequency $\omega \approx U/\hbar$ which is associated with the energy scale $U$ for "atomic" transitions [139], (or with the formation of repulsively bound pairs [137]). On the other hand, in metallic systems, the behavior for $\omega \to 0$, is in the simplest case dominated by the coherent quasiparticle contributions. Let us now discuss these contributions in correlated superlattices. For this purpose, we use the results obtained by the SBA. For the application of this method to the heterostructure model and for specific notation we refer to Chap. 4. In the following we discuss both of the above mentioned approaches (Secs. 5.2.1 and 5.2.2) to calculate the Drude weight.

## 5.2 Drude weight and optical conductivity

**Drude weight from twisted boundary conditions**

In order to evaluate (5.10) we estimate the ground-state energy and its dependence on the twist $\Phi$ using the SBA. Because the explicit dependence of the saddle-point equations [Eqs. (4.29, 4.30, 4.31)] on $\Phi$ is in second order (the ground state does not carry any current) we find that the linear change of the mean-fields in $\Phi$ vanishes

$$\left.\frac{d\,d_l}{d\Phi}\right|_{\Phi=0} = \left.\frac{d\,n_l}{d\Phi}\right|_{\Phi=0} = \left.\frac{d\,\lambda_l}{d\Phi}\right|_{\Phi=0} = 0. \tag{5.11}$$

The calculation of $D$ is thus considerably simplified because only the explicit dependence on $\Phi$ enters[1]

$$\begin{aligned} D &= \pi e^2 N_x^2 \left.\frac{\partial^2 E_G(\Phi)}{\partial \Phi^2}\right|_{\Phi=0} = -\frac{\pi e^2}{N_\parallel L} \sum_{\mathbf{k}\nu} \left[\varepsilon_{\mathbf{k}} n_{\mathbf{k}\nu}^{\text{coh}} - \frac{\partial Z_{\mathbf{k}\nu}}{\partial \varepsilon_{\mathbf{k}}}|\nabla\varepsilon_{\mathbf{k}}|^2 f_0(E_{\mathbf{k}\nu})\right] \\ &= \frac{\pi e^2}{L}\sum_\nu Z_\nu \mathcal{N}_v(\varepsilon_\nu^*) = \frac{e^2}{4\pi\hbar L a}\sum_\nu A_\nu(E_F)\bar{v}_\nu(E_F). \end{aligned} \tag{5.13}$$

For the second equality, which shows the equivalence to (5.6) most clearly, we used $n_{\mathbf{k}\nu}^{\text{coh}}$ and $Z_{\mathbf{k}\nu}$ given in Eq. (4.44) and Eq. (4.45), respectively. Furthermore, we find that

$$\frac{\partial Z_{\mathbf{k}\nu}}{\partial \varepsilon_{\mathbf{k}}} \equiv \frac{\partial^2 E_{\mathbf{k}\nu}}{\partial \varepsilon_{\mathbf{k}}^2} = 2\sum_{\nu'\neq\nu}\frac{[\sum_l z_l^2 \psi_{\mathbf{k}\nu}(l)\psi_{\mathbf{k}\nu'}(l)]^2}{E_{\mathbf{k}\nu} - E_{\mathbf{k}\nu'}} \tag{5.14}$$

which makes clear that the nonuniform renormalization of the quasiparticle dispersion is directly connected to the spatial variation of $z_l$. Integration by parts leads to the last line in (5.13) where we have introduced the weighted density of states [78]

$$\mathcal{N}_v(\varepsilon) = \frac{1}{N_\parallel a^2}\sum_{\mathbf{k}}|\nabla\varepsilon_{\mathbf{k}}|^2 \delta(\varepsilon - \varepsilon_{\mathbf{k}}) = \int\frac{d^2k}{(2\pi)^2}|\nabla\varepsilon_k|^2 \delta(\varepsilon - \varepsilon_k). \tag{5.15}$$

---

[1]The first derivative of $E_G$ with respect to $\Phi$ is

$$N_x \frac{\partial E_G}{\partial \Phi} = N_x \frac{2}{N_s}\sum_{\mathbf{k}\nu}\underbrace{\frac{\partial E_{\mathbf{k}\nu}}{\partial \varepsilon_{\mathbf{k}}}}_{=Z_{\mathbf{k}\nu}}\underbrace{\frac{\partial \varepsilon_{\mathbf{k}}(\Phi)}{\partial \Phi}}_{=\frac{1}{N_x}\partial \varepsilon_{\mathbf{k}}/\partial k_x} f_0(E_{\mathbf{k}\nu}) = \frac{1}{N_s}\sum_{\mathbf{k}\nu}|\nabla\varepsilon_{\mathbf{k}}(\Phi)|^2 \underbrace{Z_{\mathbf{k}\nu}f_0(E_{\mathbf{k}\nu})}_{=n_{\mathbf{k}\nu}^{\text{coh}}}. \tag{5.12}$$

Taking again the derivative with respect to $\Phi$ and using $\Delta\varepsilon_{\mathbf{k}} = -\varepsilon_{\mathbf{k}}$ one arrives at the right-hand side of the first line in (5.13).

For a square lattice it can be expressed in terms of complete elliptic integrals of the first and second kind:

$$\mathcal{N}_v(\varepsilon) = \frac{8t}{\pi^2}\left\{\mathrm{E}\left[1-\left(\frac{\varepsilon}{4t}\right)^2\right] - \left(\frac{\varepsilon}{4t}\right)^2 \mathrm{K}\left[1-\left(\frac{\varepsilon}{4t}\right)^2\right]\right\}$$

for $|\varepsilon| \leq 4t$ and zero otherwise. In the last expression of (5.13) we restored all the constants and used a well-known form for the Drude weight of free quasiparticles [140]: $A_\nu(E)$ denotes the area (in momentum space) of constant energy of the subband $\nu$ and

$$\bar{v}_\nu(E) = \frac{1}{A_\nu(E)}\int |\mathbf{v}_{\mathbf{k}\nu}| dA_\nu(E) \tag{5.16}$$

is the averaged velocity over the constant energy surface and we recall that the quasiparticle velocity in the SBA is found to be

$$\mathbf{v}_{\mathbf{k}\nu} = \frac{Z_{\mathbf{k}\nu}}{\hbar}\nabla\varepsilon_{\mathbf{k}}. \tag{5.17}$$

with a renormalization factor $Z_{\mathbf{k}\nu} \leq 1$ given in Eq. (4.45). In deriving (5.13) we have neglected the fact that the quasiparticle states in a superlattice of period $L$ not only depend on the 2D-momentum $\mathbf{k} \equiv (k_x, k_y)$ and the subband index $\nu$ but also on $k_\perp \equiv k_z$ with $-\pi/(La) \leq k_\perp \leq \pi/(La)$. This approximation is justified when restricting to large superlattice periods $L \gg 1$ which is implicitly assumed during the whole chapter.

**Intersubband contributions from linear response theory**

The same result (5.13) is obtained from the general linear response expression (5.6) using the free SBA quasiparticle states and replacing the kinetic energy (5.3) and the current operator (5.5) by their corresponding renormalized expressions

$$\hat{T}_x^{\mathrm{ren}} = -2t\sum_{\mathbf{k}l\sigma}\cos k_x z_l^2 f_{\mathbf{k}l\sigma}^\dagger f_{\mathbf{k}l\sigma}, \quad \hat{J}_{px}^{\mathrm{ren}} = e2t\sum_{\mathbf{k}l\sigma}\sin k_x z_l^2 f_{\mathbf{k}l\sigma}^\dagger f_{\mathbf{k}l\sigma}. \tag{5.18}$$

Moreover, due to the non-uniform spatial renormalization, the regular part of the optical conductivity, $\sigma'_{\mathrm{reg}}(\omega) = -\chi''(\omega)/\omega$, acquires non-vanishing contributions from intersubband quasiparticle scattering processes whenever $U_r > 0$:

$$\sigma'_{\mathrm{reg}}(\omega) = \frac{2\pi}{N_s}\sum_{\mathbf{k}\nu'<\nu}\left|\langle\nu\mathbf{k}|\hat{J}_{px}^{\mathrm{ren}}|\nu'k\rangle\right|^2 [f_0(E_{\mathbf{k}\nu'}) - f_0(E_{\mathbf{k}\nu})]\delta[\omega^2 - (E_{\mathbf{k}\nu} - E_{\mathbf{k}\nu'})^2]. \tag{5.19}$$

## 5.2 Drude weight and optical conductivity

The matrix elements between subband pairs are given by

$$\langle \mathbf{k}\nu|\hat{J}_{px}^{\mathrm{ren}}|\mathbf{k}\nu'\rangle = e2t\sin k_x \sum_{\sigma l} z_l^2 \psi_{\mathbf{k}\nu}(l)\psi_{\mathbf{k}\nu'}(l), \qquad (5.20)$$

which allows to evaluate (5.19). In summary, we find that the low-frequency optical conductivity consists of a Drude peak centered at $\omega = 0$ as well as a regular part which stems from intersubband transitions.

### 5.2.4 Results

In Fig. 5.2(a) we show the Drude weight $D$ [Eq. (5.13)] for superlattices with period $L = 20$ and a variable number of counterion layers $N$. We illustrate how $D$ evolves from the band insulator ($N = 0$) to the Mott insulator ($N = L$), which both have a vanishing Drude weight. The maximal $D$ as function of the averaged electronic density $N/L$ depends on the value of $E_C/t$ and shifts to lower $N$'s for increasing $E_C$ because the screening length $\lambda_{\mathrm{sc}} \sim \sqrt{t/E_C}a$ is reduced. Figure 5.2(b) shows the low-frequency optical conductivity $\sigma'(\omega)$ along with its two contributions: the Drude peak and the regular part consisting of intersubband excitations, see Eq. (5.19). We note that in superlattices with a fixed number of counterion layers $N$, the value of the "two-dimensional response" $D_{2D} \equiv DLa$ is essentially independent of $L$ for large $L$. The Drude weight can therefore be approximated by $D = D_{2D}/(La)$ where $D_{2D}$ is obtained from the two-dimensional response of the quantum-well model discussed in Chap. 4.

Figure 5.3(a) shows a comparison of the calculated oscillator strength $D \approx D_{2D}/(La)$ with the results of optical experiments reported by Seo et al. in Ref. [111] on $(\mathrm{LaTiO_3})_N/(\mathrm{SrTiO_3})_{M=10}$ superlattices. The overall variation of $D$ with the averaged charge $N/(N+M)$ is only consistent with the experimental data in the strongly correlated Mott regime $U_r > U_c \approx 16t$.[2] From this comparison we obtain an order of magnitude estimate for the hopping matrix element: $t \approx 0.15$ eV. In contrast, in the weakly interacting Hartree regime $U_r \ll U_c$, the qualitative behavior of $D$ as function of $N/(N+M)$ is quite different and clearly not consistent with the experimental data.[3] On the other hand, even in

---

[2] For a characterization of Hartree and Mott regime see Sec. 4.4.1.

[3] However, we have to keep in mind that the possibility of a magnetic order in the central layers, which is expected to reduce the variation of $D$ with $N$, is ignored in our considerations.

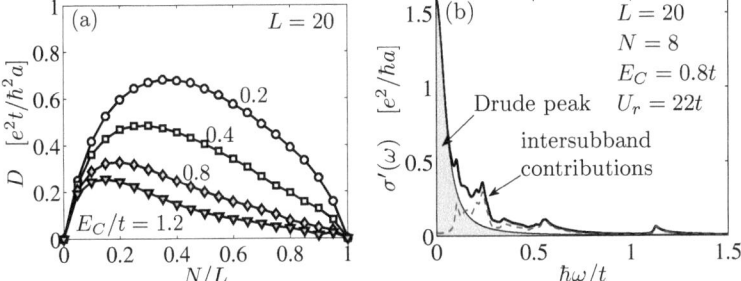

Figure 5.2: (a) The Drude weight $D$ as function of $N/L$ for different values of $E_C$ for a given superlattice period $L = 20$. Both the pure band insulator ($N = 0$) and the pure Mott insulator ($N = L$) have a vanishing $D$. (b) The low-frequency optical conductivity of the $N = 8$ superlattice showing the Drude peak (broadened by $0.05t$) and the regular part containing intersubband contributions. In both panels, the value of the onsite interaction is fixed at $U_r = 22t$.

the strongly correlated limit the calculations reveal a characteristic dependence of $D$ on $N/L$: initially, $D$ increases with $N$, shows a maximum at $N = 3 - 4$ and then decreases again. The increase of $D$ up to $N \sim \lambda_{sc}$ is expected for a system with dominant contributions from the interface region. The subsequent decrease is more surprising since in this situation one would expect saturation with $N$ at $N \approx 2\lambda_{sc}/a$. To illuminate this point we consider in Fig. 5.3(b) the two dimensional response $D_{2D}$ along with its two contributions according to Eq. (5.6). Indeed, contrary to the expectations, we find that the two-dimensional value $D_{2D}$ saturates at much larger $N$'s ($N \geq 20$) indicating that the spatial coherence of the quasiparticle states at the Fermi energy extends over a much larger region then just the interface region, see also Sec. 4.5 on the interfacial heavy-fermion scenario. On the other hand, the regular part of the optical conductivity obtains more and more weight at ever lower frequencies when increasing $N$. This spectral weight is proportional to $-\chi'_{2D}(0)$ and is also shown in Fig. 5.3(b). As a matter of fact, the $f$-sum rule (5.7) dictates that the integrated low-frequency optical conductivity is proportional to the kinetic energy of the coherent quasiparticles

## 5.3 Thermoelectricity

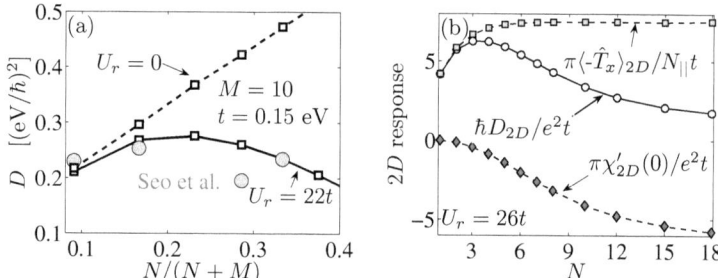

Figure 5.3: (a) The Drude weight $D$ of $(MI)_N/(BI)_{M=10}$ superlattices, as found by the present theory for $t = 0.15$ eV, and $U_r = 22t$ (solid line) and $U_r = 0$ (dashed line). The filled circles denote the experimental data of Seo et al. on $(LaTiO_3)_N/(SrTiO_3)_{M=10}$ superlattices reported in Ref. [111]. (b) The two-dimensional response $D_{2D} = DLa$ as function of $N$ in the strongly correlated regime along with its two contributions according to Eq. (5.6).

$\langle -\hat{T}_x \rangle_{2D}$ and it is this quantity which saturates rapidly at $N \approx 5$ as expected.

## 5.3 Thermoelectricity

The electronic structure which was discussed in Chap. 4 shows features reminiscent of heavy-fermion physics (the hybridization of itinerant and localized degrees of freedom in the interface region) which may provide favorable conditions to enhance the low-temperature thermoelectrical properties [70, 71]. We discuss these aspects using a semiclassical point of view to obtain the generalized transport coefficients. We thus assume diffusive transport but include the quantization of the electronic structure into subbands. On the other hand, thermopower is also connected to entropy contributions which can be increased through the presence of localized (orbital and spin) degrees of freedom. This aspect will be discussed using a generalized Heikes formula to access the high-temperature regime. Phenomenological aspects related to the thermoelectrical response are discussed in

App. B.1.

### 5.3.1 Generalized transport coefficients

The self-consistently renormalized quasiparticle subbands at $T = 0$ were taken to calculate the transport properties of the metallic electron system in the superlattice structure. This restricts our analysis to lowest order in $T$.

**Relaxation time approximation**

We use the Boltzmann transport theory in the relaxation time approximation to obtain the generalized transport coefficients. We neglect the (residual) interactions between Landau's Fermi-liquid quasiparticles [1]. The relation to the linear-response Kubo formula is discussed in App. B.2. We introduce the local quasiparticle distribution function $f_{\mathbf{k}\nu}$ for quasiparticles in the subbands $\nu$ with two-dimensional momentum $\mathbf{k}$ (we suppress the spin index $\sigma$). The linear response to an applied uniform electric (electrochemical) field $\mathbf{E}$ or temperature gradient $\nabla T$ in the direction perpendicular to the growth direction of the superlattice ($xy$-plane) is found by linearizing the distribution function $f_{\mathbf{k}\nu} = f_T(E_{\mathbf{k}\nu} - \mu) + g_{\mathbf{k}\nu}$, where $f_T(E) = [1 + \exp(\beta E)]^{-1}$ is the equilibrium Fermi-Dirac distribution function for the inverse temperature $\beta = 1/k_B T$ and where $g_{\mathbf{k}\nu}$ is proportional to the applied field.

For the setup shown in Fig. 5.1 it is sufficient to restrict the calculations to the $x$ component of the currents. For this case we define the *transport distribution function* $\Phi(E)$ as follows [141]:

$$\Phi(E) = \frac{e^2}{4\pi^2 \hbar L a} \sum_\nu \tau_\nu(E) \bar{v}_\nu(E) A_\nu(E) \equiv \sum_\nu \tau_\nu(E) D_\nu(E). \quad (5.21)$$

Here, we introduced the relaxation time $\tau_\nu(E)$ which is assumed to be independent of the two-dimensional momentum $\mathbf{k}$. As before, $A_\nu(E)$ denotes the area of the constant energy surface at energy $E$ and the averaged quasiparticle velocity $\bar{v}_\nu(E)$ of the subband $\nu$ is given in Eq. (5.16). According to Eq. (5.21) the transport distribution function is a sum of the contributions of each partially filled subband,

$$\Phi(E) = \sum_\nu \Phi_\nu(E), \quad (5.22)$$

## 5.3 Thermoelectricity

which allows us to study the different contributions to the transport coefficients. We define the following Fermi integrals over the single kernel function $\Phi(E)$,

$$L^{(\alpha)} = \int dE \left(-\frac{\partial f^0}{\partial E}\right) E^\alpha \Phi(E+\mu). \tag{5.23}$$

For the conductivity $\sigma$, the thermopower or Seebeck coefficient $S$, and the thermal conductivity of the electrons $\kappa^e$, one then finds [140, 141]

$$\sigma = L^{(0)}, \tag{5.24}$$

$$S = -\frac{1}{eT}\frac{L^{(1)}}{L^{(0)}}, \tag{5.25}$$

and

$$\kappa^e = \frac{1}{e^2 T}\left[L^{(2)} - \frac{\left(L^{(1)}\right)^2}{L^{(0)}}\right]. \tag{5.26}$$

### Characterization of the electronic structure

At low temperatures we expect that important aspects of transport can be understood by the properties of the quasiparticles. We therefore consider the effect of the electron-electron interaction as reflected in the renormalization of the quasiparticle velocity of the subband $\nu$,

$$\mathbf{v}_{\mathbf{k}\nu} = \left(\frac{\partial E_{\mathbf{k}\nu}}{\partial \varepsilon_{\mathbf{k}}}\right)\frac{1}{\hbar}\nabla \varepsilon_{\mathbf{k}}, \tag{5.27}$$

and its energy dependence,

$$\alpha_{\mathbf{k}\nu} = \Delta\varepsilon_{\mathbf{k}}\frac{\partial}{\partial \varepsilon_{\mathbf{k}}}\log\left(\frac{\partial E_{\mathbf{k}\nu}}{\partial \varepsilon_{\mathbf{k}}}\right), \tag{5.28}$$

where we have defined $\Delta\varepsilon_{\mathbf{k}} = \varepsilon_{\mathbf{k}} - \varepsilon_b$ with $\varepsilon_b = -4t$ the energy of the lower band edge. As discussed in Chap. 4, in the slave-boson approximation, the inverse effective mass is equal to the quasiparticle weight and we therefore use the notation $Z_{\mathbf{k}\nu} = \partial E_{\mathbf{k}\nu}/\partial \varepsilon_{\mathbf{k}}$ to denote the renormalization factor of the velocity. The dimensionless parameter $\alpha_{\mathbf{k}\nu}$ quantifies a particle-hole asymmetry induced by strong correlations and is important for the thermopower. In Fig. 5.4(a) we show the quasiparticle dispersion $E_{\mathbf{k}\nu}$ for the $N = 15$, $M = 5$ superlattice. The value of

*Optical conductivity and thermoelectricity in correlated superlattices*

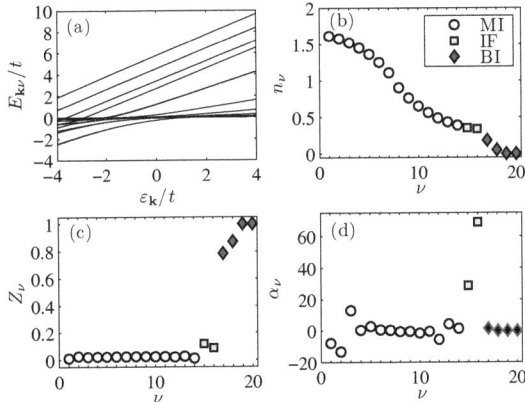

Figure 5.4: (a) The quasiparticle dispersion $E_{\mathbf{k}\nu}$ as function of the non-interacting in-plane dispersion $\varepsilon_{\mathbf{k}}$. (b) The subband filling $n_\nu$, (c) the quasiparticle weight $Z_\nu$ and (d) the induced particle-hole asymmetry $\alpha_\nu$ for the individual subbands which correspond to the dispersion in (a). Circles are associated with the subbands of the MI region, squares with the IF and diamonds with the BI region. The superlattice parameters are $N = 15$, $M = 5$, $U_r = 22t$ and $E_C = 0.8t$.

the onsite repulsion, $U_r = 22t$, is well above the critical interaction strength of the Mott transition in the half-filled bulk system,[4] $U_c \approx 16t$. Thus, the quasiparticle dispersion shown in Fig. 5.4 corresponds to the strongly correlated regime. In panels (b) to (d) we show the subband filling $n_\nu$, the quasi-particle weight $Z_\nu$ and the induced particle-hole asymmetry $\alpha_\nu$ at the Fermi energy for the subbands shown in (a). As shown below, these quantities are important ingredients for determining the quasiparticle transport of the *interfacial heavy-fermion state* obtained by the SBA, see Sec. 4.5.

---

[4] See Sec. 3.2 for more details.

## 5.3 Thermoelectricity

**Reduction of complexity**

As it is apparent from Fig. 5.4 we usually have to deal with a whole bunch of partially filled subbands. For a qualitative discussion of the generalized transport coefficients it is useful to group the different subband states according to the regions where most of the spatial weight of their wave function is located. We thus define $\nu_{\mathrm{MI}} = 1, \ldots, N-1$, $\nu_{\mathrm{IF}} = N, N+1$ and $\nu_{\mathrm{BI}} = N+2, \ldots, L$. Indeed, as shown in panels (b) to (d), the quasiparticle properties associated with this three regions are very similar. For example, note that due to strong local correlations, the quasiparticle weights $Z_{\nu_{\mathrm{MI}}}$ and $Z_{\nu_{\mathrm{IF}}}$ are strongly reduced for the subbands of the MI and IF region [panel (c)] whereas the particle-hole asymmetry $\alpha_{\nu_{\mathrm{IF}}}$ is enhanced most dominantly for the subbands of the IF region [panel (d)]. Formally, we can introduce for each region a transport distribution function

$$\Phi(E) = \sum_{n=\mathrm{BI,IF,MI}} \Phi_n(E), \quad \Phi_n(E) = \sum_{\nu_n} \Phi_\nu(E) \equiv \tau_n(E) D_n(E) \qquad (5.29)$$

where we have introduced the "Drude weight" $D_n(E)$ and the (averaged) relaxation time $\tau_n(E)$ of region $n$:

$$D_n(E) = \sum_{\nu_n} D_\nu(E), \quad \tau_n(E) = \sum_{\nu_n} \tau_\nu(E) \frac{D_\nu(E)}{D_n(E)}, \quad n = \mathrm{BI, IF, MI}. \qquad (5.30)$$

### 5.3.2 Estimation of the relaxation time

We start with an estimate for the relaxation time obtained by a microscopic consideration of elastic scattering off impurities or vacancies. For simplicity, we focus on the short-range impurity Hamiltonian

$$V_{\mathrm{imp}} = V_0 \sum_{\sigma i'} c^\dagger_{i'\sigma} c_{i'\sigma}, \qquad (5.31)$$

where $i'$ labels the lattice sites of the impurities. We neglect multiple-scattering by different impurities and use the single-site $T$-matrix approximation [142] with an average over the locations $i$ of the impurity. This average introduces a sum over layers $l$ and we find

$$\left\langle \left| T^{(i)}_{\nu\nu'}(E_{\mathbf{k}\nu}) \right|^2 \right\rangle_{\mathrm{av}} = \sum_l \left| \frac{\psi_{\mathbf{k}\nu}(l) V_0 \psi_{\mathbf{k}'\nu'}(l)}{1 - V_0 G^{\mathrm{QP}}_{ll,\sigma}(E_{\mathbf{k}\nu})} \right|^2. \qquad (5.32)$$

Here, $T^{(i)}_{\nu\nu'}$ is the atomic $T$-matrix for a single impurity at site $i$ and

$$G^{\text{QP}}_{ll,\sigma}(E) = \sum_\nu \int \frac{d^2k}{(2\pi)^2} \frac{\psi_{\mathbf{k}\nu}(l)^2}{E - E_{\mathbf{k}\nu} + i0^+} \tag{5.33}$$

is the retarded local quasiparticle Green's function of the pure system. We neglect the real part of $G^{\text{QP}}_{ll\sigma}$ and introduce the layer-resolved quasiparticle density of states

$$\rho^*_{l,\sigma}(E) = -\frac{1}{\pi}\text{Im}\, G^{\text{QP}}_{ll,\sigma}(E) = \sum_\nu \int \frac{d^2k}{(2\pi)^2} \psi_{\mathbf{k}\nu}(l)^2 \delta(E - E_{\mathbf{k}\nu}). \tag{5.34}$$

The scattering rate follows as

$$\frac{1}{\tau_\nu(E)} = \frac{2\pi}{\hbar} V_0^2 c_{\text{imp}} \sum_l \frac{\psi_{E\nu}(l)^2 \rho^*_{l,\sigma}(E)}{1 + \pi^2 V_0^2 \rho^*_{l,\sigma}(E)^2}, \tag{5.35}$$

where $c_{\text{imp}}$ is the impurity concentration assumed to be independent of the layer $l$. Using the relation (5.30) we arrive at the averaged scattering rate $1/\tau_n$ associated with the three regions. Figure 5.5 shows $1/\tau_n(E_F)$ evaluated at the Fermi energy $E_F$ as function of the impurity potential $V_0$. For comparison we give in the inset the result obtained for all the partially filled subbands. Clearly, the influence of $V_0$ is different in the different regions because of the spatial non-uniformity of the system: the value of $\rho^*_{l,\sigma}(E_F)$ strongly changes with $l$ reaching its maximum in the center of the heterostructure where the quasiparticle states are confined to energies near the Fermi energy $E = 0$ which results in a strong variation of the scattering rates for the different subbands. The first Born approximation is recovered in the limit of weak impurities: $\pi V_0 \rho^*_{l,\sigma}(E_F) \ll 1$ for all $l$. In this case, $1/\tau_n(E_F)$ is substantially bigger for the subbands of the MI and IF region. In other words, the slower quasiparticles are scattered more strongly. In this case, the quasiparticle transport is mainly determined by the properties of the almost empty subbands of the BI region. In the unitary limit, $\pi V_0 \rho^*_{l,\sigma}(E_F) \gg 1$, the scattering rate of the slower quasiparticles is significantly reduced. In this case, the quasiparticle transport is mainly determined by the strongly renormalized subbands of the MI and IF region. For moderate values of $V_0$, the scattering rates are comparable.

## 5.3 Thermoelectricity

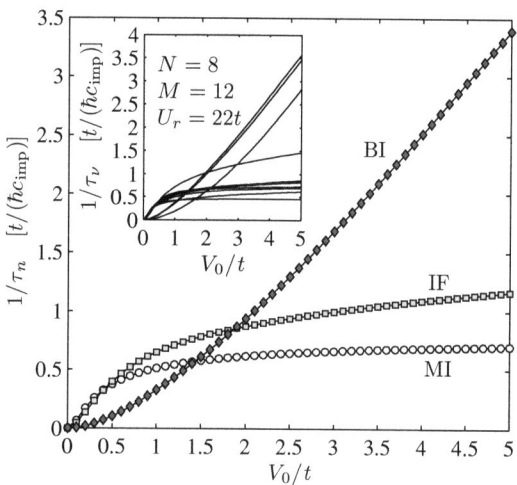

Figure 5.5: The averaged scattering rates $1/\tau_n$ evaluated at the Fermi energy for the subbands associated with the three different regions MI, IF, BI as function of the impurity strength $V_0$. The inset shows the scattering rate of all the partially filled subbands.

From the above example we expect that in general the dependence of the scattering rate on microscopic parameters is different for the different subbands but that it is reasonable to group them according to the three spatial regions. Beside the impurity potential strength, there are other parameters that can influence the subband dependence of $\tau_\nu$ (or $\tau_n$) in real systems. For example, it is conceivable that the scattering occurs mainly due to imperfections of the interface in which case $1/\tau_{\text{IF}}$ is largest. Furthermore, there is good reason to believe that at finite temperatures electron-electron or electron-phonon scattering will affect the individual subbands in a different way but that it is still reasonable to consider three different groups (BI, IF and MI). We therefore think that for a qualitative discussion of many aspects it is sufficient to keep three different relaxation times ($\tau_{\text{BI}}$, $\tau_{\text{IF}}$ and $\tau_{\text{MI}}$) which we can consider as phenomenological parameters. A further simplification can then be obtained by assuming energy independent relaxation times $\tau_n$.

### 5.3.3 DC conductivities

For completeness we explicitly state here the expressions for the electrical and thermal conductivities obtained in the relaxation time approximation [140, 143]. From Eq. (5.24) it follows that at low temperatures the dc conductivity is constant and given by

$$\sigma = \Phi(E_F) = \frac{e^2}{4\pi^2 \hbar L} A_F \Lambda. \tag{5.36}$$

$A_F = \sum_\nu A_\nu(E_F)$ is the total area of the Fermi surface and the mean free path is obtained as a weighted average,

$$\Lambda = \frac{1}{A_F} \sum_\nu A_\nu(E_F) \bar{v}_\nu(E_F) \tau_\nu(E_F). \tag{5.37}$$

We note here that in optical measurements on $(\text{LaTiO}_3)_N/(\text{SrTiO}_3)_{M=10}$ superlattices [111], a value of $\tau \approx 35$ fs for the relaxation time at $T = 10$ K was found from which an averaged mean-free path of $\Lambda \approx 10$ nm $\approx 25\,a$ has been deduced. The electronic contribution to the thermal conductivity follows from Eq. (5.26),

$$\kappa_{2D}^e = \frac{\pi^2}{3} \frac{k_B^2 T}{e^2} \Phi(E_F), \tag{5.38}$$

and is related to the electrical conductivity by the Wiedemann-Franz law typical for elastic scattering [143].

### 5.3.4 Thermopower

Let us know consider the thermopower $S$, which, in an open circuit, is defined as the constant between electrochemical field and temperature gradient $\mathbf{E} = S\nabla T$. For a metal at low temperatures, it follows from Eq. (5.25) that $S$ is given by the Mott formula [143]

$$S = -\frac{\pi^2}{3} \frac{k_B^2 T}{e} \frac{\partial}{\partial E} \log \Phi(E) \bigg|_{E=E_F}. \tag{5.39}$$

## 5.3 Thermoelectricity

It follows from (5.21) that in the superlattice system considered here, we can interpret the thermopower as a weighted sum of the contributions of the partially filled quasiparticle subbands,

$$S = \sum_\nu \frac{S_\nu \sigma_\nu}{\sigma} \qquad (5.40)$$

where $\sigma_\nu$ and $S_\nu$ are the conductivity and the thermopower of the subband $\nu$, respectively. Alternatively, we again combine the subbands into three groups $n =$ BI, IF, MI and write

$$S = \sum_n \frac{S_n \sigma_n}{\sigma}. \qquad (5.41)$$

where $S_n$ and $\sigma_n$ are defined by the corresponding transport distribution function $\Phi_n(E)$, see Eq. (5.29). As discussed above, this point of view is natural if the scattering mechanisms are similar within a unit of the subbands.

### Implications from presence of multiple subbands

From the general form (5.40) for a system which contains multiple (sub)bands it follows that $|S|$ is bounded by its maximal contribution,

$$|S| \leq \max_\nu |S_\nu|, \qquad (5.42)$$

and is in general less due to possible cancellations in the enumerator of Eq. (5.40) between electron-like ($S_\nu < 0$) and hole-like ($S_\nu > 0$) contributions. As a consequence, in a rigid-band picture (meaning that the dispersion $E_{\mathbf{k}\nu}$ is independent of the filling of the band), a multi-band system is always less efficient compared to the single-band system with the best properties. However, this conclusion breaks down in strongly-correlated electron systems because the rigid-band picture is in general not applicable and therefore a multi-band system can also yield rather large values when individual bands show an enhanced particle-hole asymmetry. Indeed, we show below that strong local correlations are favorable. They are strongest near half filling which, on the other hand, leads to several overlapping subbands in superlattice structures. Nevertheless, also in this situation we have to expect that partial cancellation will reduce the value of the total Seebeck coefficient.

**Subband contributions in the low-temperature limit**

In lowest order in the temperature we find for the contribution of an individual subband

$$S_\nu = -\frac{\pi^2}{3}\frac{k_B}{e}\frac{k_B T}{Z_\nu \Delta\varepsilon_\nu^*}\left[\Delta\varepsilon_\nu^*\left(\frac{\tau_\nu'}{\tau_\nu}+\frac{\mathcal{N}_\nu'}{\mathcal{N}_\nu}\right)+\alpha_\nu\right] \quad (5.43)$$

where $\Delta\varepsilon_\nu^* = \varepsilon_\nu^* - \varepsilon_b$ is measured from the band edge and defined through $E_\nu(\varepsilon_\nu^*) = E_F$, $\tau_\nu(\varepsilon)$ is the relaxation time of the subband $\nu$ and the prime denotes the derivative with respect to $\varepsilon$ at $E_F$. There is an overall reduction of the energy scale by $Z_\nu$. As a result, the low-temperature slope of $S_\nu$ is enhanced by a factor of $1/Z_\nu$ due to correlation effects. Such a renormalization of the Seebeck coefficient is also known to occur in the Fermi-liquid regime of the homogeneous Hubbard model [144] and reflects the reduction of the effective coherence temperature. In the following we will discuss the different contributions to $S_\nu$. The first term inside the square brackets of Eq. (5.43) describes the influence of the scattering process. Usually, when considering elastic impurity scattering, this term is small and can be neglected. The second term describes the contribution from the uncorrelated band structure. It is sizable if the subband occupation $n_\nu$ is small, because the particle-hole asymmetry is large near the subband edge. In fact, for the almost empty subbands with $n_\nu a^2 \ll 1$, we find

$$\frac{\mathcal{N}_\nu'(\varepsilon_\nu^*)}{\mathcal{N}_\nu(\varepsilon_\nu^*)} \approx \frac{1}{4t+\varepsilon_\nu^*} = \frac{1}{2t\pi n_\nu a^2}. \quad (5.44)$$

On the other hand, in the presence of strong electron correlations, the third term $\alpha_\nu$ can give a dominant contribution. As mentioned above, the dimensionless parameter $\alpha_\nu$ quantifies a particle-hole asymmetry induced by strong local interaction in the inhomogeneous system and is largest for the subbands of the interface region, see Fig. 5.4(d). We have found that these contributions can lead to a substantial enhancement of the total thermopower.

### 5.3.5 Dependence of the thermopower on model parameters

We have studied the dependence of the low-temperature thermopower on the model parameters $U_r$, $E_C$, and $N$. In summary, we find that the interface contribution to the thermopower is enhanced (on the negative side) by

## 5.3 Thermoelectricity

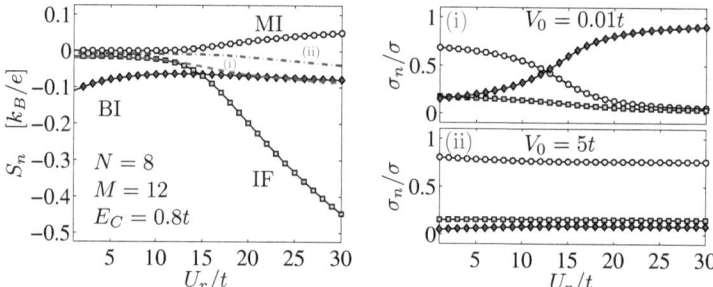

Figure 5.6: Left. The contribution $S_n$ to the total thermopower $S$ at $k_B T = 0.01t$ from the three different regions as function of $U_r$. These contributions are weighted with their relative conductivity $\sigma_n/\sigma$ shown for two cases: (i) for a impurity potential $V_0 = 0.01t$ (Born limit) and (ii) for $V_0 = 5t$ (unitary limit). Right. The relative conductivity $\sigma_n/\sigma$ for (i) and (ii).

(i) strong onsite interactions $U_r > U_c$,

(ii) large values of $N$ such that bulk-like properties in the center of the MI are obtained and

(iii) a sharp interface (short screening length) $E_C > t$.

To which extent also the total thermopower is enhanced in the above limits depends, according to the relation (5.41), on the relative conductivity $\sigma_n/\sigma$ and therefore on $\tau_n$ in the relaxation time approximation. In all cases (i) to (iii), the observed enhancement of the interface contribution can be associated with a reduction of the effective hybridization between the interface states and the localized degrees of freedom of the MI region which leads to a large particle-hole asymmetry $\alpha_\nu$. In the following we give two illustrative examples.

Figure 5.6 shows the dependence of $S_n$ on $U_r$ for a superlattice with $N = 8$ and $M = 12$. While the variation of the BI and MI contribution is rather weak, $|S_\text{IF}|$ shows a clear enhancement due to the increasing particle-hole asymmetry. The total thermopower $S$ depends on $\sigma_n/\sigma$ and we show the results for two different

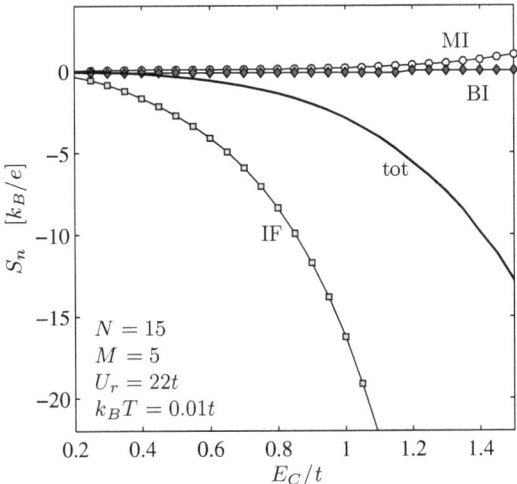

Figure 5.7: The contribution $S_n$ to the total thermopower $S$ at $k_B T = 0.01t$ from the three different regions as function of $E_C$. For the total thermopower $S$ a constant relaxation time is assumed $\tau_{\text{MI}} = \tau_{\text{IF}} = \tau_{\text{BI}}$.

cases obtained in the T-matrix approximation: (i) weak impurity potential $V_0 = 0.01t$ (Born limit) and (ii) strong impurity potential $V_0 = 5t$ (unitary limit). In both cases, the enhancement of $S$ is much less pronounced compared to $S_{\text{IF}}$. Nevertheless, $S$ follows the general trend of $S_{\text{IF}}$.

As a second example we show in Fig. 5.7 the dependence of the thermopower on $E_C$ for a superlattice with $N = 15$ in the strongly correlated regime $U_r = 22t$. We find that the sharper the interface (the bigger $E_C/t$) the more is $|S_{\text{IF}}|$ enhanced. Again this can be understood by the fact that the particle-hole asymmetry parameter $\alpha_\nu$ is large for the subbands of the interface region in the limit of a sharp interface. In addition, we show also the total thermopower $S$ as obtained by assuming $\tau_{\text{MI}} = \tau_{\text{IF}} = \tau_{\text{BI}}$.

## 5.3 Thermoelectricity

### 5.3.6 Quantum oscillations in the thermopower

There is another interesting behavior of the thermopower which is a consequence of the spatial confinement of the conduction electrons. Namely, in the case where the contributions from the MI region are dominant,[5] the quantum mechanical confinement manifests itself in an oscillatory behavior of $S$ as function of $N$. Indeed, we observe that for $N$ even, $S$ can take positive values at low temperatures because there is one subband close to half filling with a dominant positive contribution. As a consequence, $S$ oscillates as function of $N$ (see Fig. 5.8). Moreover, the larger the $N$ the larger the effect of cancellation becomes, which reduces the amplitude of the oscillation for $N \to \infty$. In Fig. 5.8 the Seebeck coefficient is shown for different $N$ in a quantum well geometry for a strong impurity scattering potential $V_0 = 10t$. The value of the low-temperature thermopower $S$ shows an even-odd oscillation with a decreasing amplitude for $N \to \infty$.

### 5.3.7 Thermoelectrical figure of merit and power factor

For thermoelectric applications, the dimensionless figure of merit

$$Z_T T = \sigma S^2 T / (\kappa^e + \kappa^L) \qquad (5.45)$$

should be as large as possible. This quantity provides a measure of the efficiency of a material used for cooling or heating, respectively, as a thermoelectric power generator, see App. B.1. It involves the static transport coefficients of the system, the electrical conductivity $\sigma$, the thermoelectric power or Seebeck coefficient $S$ and the total thermal conductivity $\kappa = \kappa^e + \kappa^L$. A high value of $\sigma$ is necessary to minimize Joule heating, while a low value of $\kappa$ helps to maintain a large temperature gradient. In metal-oxide based thermoelectric materials, the phonon thermal conductivity usually plays a dominant role and often $\kappa^L \gg \kappa^e$ (see, e.g., Ref. [125] and [134]). Therefore, one often focuses on a quantity involving only the electronic transport properties, namely the product PF$= \sigma S^2$ called the power factor. It is a common approach to optimize PF by varying geometrical parameters (e.g. $N$ and $M$) of the superlattice [133, 145]. Calculating PF for various geometries we find that PF is bigger the bigger $N$, similar to $S$. However,

---

[5] In the $T$-matrix approximation, this corresponds to a strong impurity potential $V_0$, as seen from the behavior of $1/\tau_n$ in Fig. 5.5

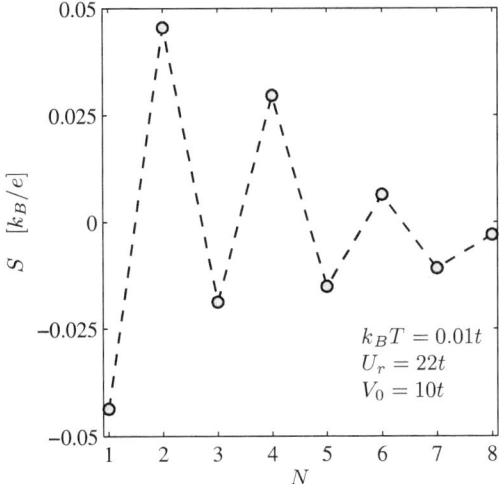

Figure 5.8: Quantum oscillations in the thermopower $S$ at $k_B T = 0.01t$ as function of $N = 1 - 8$ in the limit of a strong impurity potential $V_0 = 10t$.

this conclusion is only true in lowest order in the temperature and we expect that when taking into account higher order terms it is possible to optimize PF. The problem is that the SBA restricts our analysis to lowest order, see above, but when courageously extended to finite temperatures we have indeed found the possibility to optimize PF by varying $N$ for fixed $M$ [70].

### 5.3.8 Thermopower in the atomic limit

The thermopower is also connected to the entropy of the system which can be increased through the presence of localized (orbital and spin) degrees of freedom. Recently, it was proposed that these contributions are important to explain the large thermopower observed in $NaCo_2O_4$ [132, 146]. The contribution from the

## 5.3 Thermoelectricity

entropy in the high-temperature limit is in general referred to as 'Heikes formula' and in the following we consider a generalization to strongly correlated superlattices.

In the *homogeneous* system it follows from the Kubo formula that the thermopower $S$ in the high temperature limit is given by [49, 128]

$$S = \frac{\mu}{eT}. \tag{5.46}$$

But from Maxwell's relations this is related to the entropy S per particle (at fixed energy $E$ and volume $V$)

$$\frac{\mu}{T} = -\left(\frac{\partial S}{\partial N}\right)_{V,E}. \tag{5.47}$$

The evaluation of the right hand side yields Heikes formula. Let us now discuss the situation $t \ll k_B T \ll U$ such that doubly occupied sites are completely suppressed. Then, using Eq. (2.22), we find at density $n$,

$$S = -\frac{k_B}{e}\log\left[\frac{2(1-n)}{n}\right] \tag{5.48}$$

where the log 2 contribution arises from the entropy of the spin degree of freedom. As pointed out in Ref. [132], in transition metal oxides also the inclusion of orbital degrees of freedom is necessary. This is in principle straight forward but we will not discuss it further here. For the *inhomogeneous* system an appropriate generalization of Eq. (5.48) is given by

$$S = \frac{\sum_l S_l \sigma_l}{\sigma} = -\frac{k_B}{e}\frac{\sum_l \log\left[\frac{2(1-n_l)}{n_l}\right] n_l(1-n_l)}{\sum_l n_l(1-n_l)}. \tag{5.49}$$

Here we have used the *local* quantities

$$S_l = -\frac{k_B}{e}\log\left[\frac{2(1-n_l)}{n_l}\right], \quad \sigma_l = \frac{e^2 A\beta}{2} n_l(1-n_l), \tag{5.50}$$

with a temperature and doping independent constant $A$ [147]. The derivation of Eq. (5.49) is again based on the Kubo formalism to the transport coefficients and is outlined in the App. B.2.3. The weighted sum in Eq. (5.49) clearly shows that an inhomogeneous system is not favorable as long as a *local* description is appropriate. In fact, if one seeks to optimize the powerfactor PF= $S^2\sigma$ in this limit for a spatially varying density profile, the optimal solution is found to be the homogeneous solution with optimal density $n \approx 0.12$, c.f. Ref. [147]. This is exactly the opposite behavior than found in the low-temperaure limit discussed in Sec. 5.3.5. Restricting to purely electronic contribution, we therefore conclude that a spatially non-uniform system can only be favorable if spatial coherence is sustained.

## 5.4 Conclusions

In this chapter we have focused on the metallic properties in Mott-insulator/band-insulator heterostructures. We propose an analysis based on a quasiparticle picture that is expected to be valid in the Fermi liquid regime below the coherence temperature. A self-consistent renormalization of the quasiparticles was obtained by applying the slave-boson mean-field approximation and the quasiparticle transport was calculated in the relaxation-time approximation of the Boltzmann transport equation.

We have presented two consistent ways to calculate the Drude weight in the low-frequency optical conductivity. In the strongly correlated regime, the response is dominated by the contributions from the interface. This is consistent with previously published spectroscopic data. Furthermore, the comparison with experiment allows to estimate some of the key parameters of the model.

We have shown that correlation effects at low temperatures can enhance the thermoelectric response as compared to a metal/ band-insulator interface. The favorable conditions arise due to the hybridization of itinerant and localized degrees of freedom in the interface region. On the level of the model parameters this effect is increased by increasing $U$, $E_C$ and the width $N$ of the Mott insulating material confined in the heterostructure. These observations are contrasted to the results obtained by a generalized Heikes formula for strongly correlated inhomogeneous systems valid in the high temperature limit. The loss of quantum mechanical coherence extending over several atomic monolayers near the interface makes an inhomogeneous electronic density profile less favorable compared to the uniform case.

Whether the thermoelectric properties of a superlattice of the kind considered

## 5.4 Conclusions

in this chapter can overtop those of a bulk solid solution with the same composite remains unclear and further studies are necessary to clarify this point. The thermoelectric response, in particular the Seebeck coefficient, is rather sensitive to the considered scattering mechanism. Even when focusing on $s$-wave impurity scattering at low temperatures, two completely different low-temperature behavior of $S$ can be obtained in the Born and the unitary limit. Especially, this difference is manifest when considering $S$ as function of the width $N$ of the quantum well. On the other hand, measurements of the thermopower would actually be a good probe to test to which extent the quasiparticle description holds and, where applicable, to gain information on the delicate point of which subband contributions are most dominant for the quasiparticle transport.

# Appendix A

# for chapter 3

## A.1 Pseudo-spin wave analysis

In this appendix we give some technical details concerning the spin-wave analysis of Sec. 3.3.5.

### A.1.1 Canonically transformed Hamiltonian

To start we represent the pseudospin operators in terms of the canonically transformed Schwinger-Wigner bosons $a_i^{(\dagger)}$ and $b_i^{(\dagger)}$, see Eq. (3.50) and (3.51). For $I_i^z$ we find

$$I_i^z + \frac{1}{2} \equiv y_i^\dagger y_i = 2d^2 + (1-4d^2)b_i^\dagger b_i + \sqrt{2}d\sqrt{1-2d^2}(a_i^\dagger b_i + b_i^\dagger a_i), \qquad (A.1)$$

and similarly for $2I_i^x$

$$2I_i^x \equiv x_i^\dagger y_i + y_i^\dagger x_i = 2\sqrt{2}d\sqrt{1-2d^2}(1-2b_i^\dagger b_i) + (1-4d^2)(a_i^\dagger b_i + b_i^\dagger a_i). \qquad (A.2)$$

Using the representations (A.1) and (A.2) for the pseudospin operators we arrive at the representation $H_B(d)$ of the original pseudospin Hamiltonian (3.44).

### A.1.2 Effective Hamiltonian for excitations

The derivation of the effective Hamiltonian describing excitations and fluctuations around the classical ground state (3.52) consists of three steps [148]:

## A.1 Pseudo-spin wave analysis

(i) First we expand $H_B(d)$ up to second order in the $b$ bosons. This requires the expansion of the hopping term. Using (A.2) and dropping terms of order $\mathcal{O}(b^3)$ and higher we find

$$\begin{aligned}
4I_i^x I_j^x &\approx 8d^2(1-2d^2) - 16d^2(1-2d^2)(b_i^\dagger b_i + b_j^\dagger b_j - 2b_i^\dagger b_i b_j^\dagger b_j) \\
&+ 2\sqrt{2}d\sqrt{1-2d^2}(1-4d^2)(a_i^\dagger b_i + b_i^\dagger a_i + a_j^\dagger b_j + b_j^\dagger a_j) \\
&+ (1-4d^2)^2(a_i^\dagger a_j^\dagger b_i b_j + a_i a_j b_i^\dagger b_j^\dagger + a_i^\dagger a_j b_j^\dagger b_i + a_j^\dagger a_i b_i^\dagger b_j).
\end{aligned}$$

(ii) The parameter of the unitary transformation, i.e. $d$, is now determined by the requirement that the quadratic mixing terms in $a$ and $b$ of $H_B(d)$ vanish. For the common pre-factor of these terms we find

$$-\frac{U_c}{2}N\sqrt{2}d\sqrt{1-2d^2}\left(1-4d^2-u\right)\sum_i(\ldots) \quad \Rightarrow \quad d^2 = \begin{cases} \frac{1}{4}(1-u) & u \leq 1 \\ 0 & u > 1 \end{cases},$$

which yields the result of the Gutzwiller approximation for the double occupancy density, (3.22).

(iii) The next step is to replace the $a$ operators by the complex number $a_0$ which is the condensate amplitude that we set equal to 1. This is equivalent to the statement that fluctuations around the classical ground state are assumed to be small. In this way, an effective Bogoliubov Hamiltonian for the $b$ operators is obtained, namely, for $u \leq 1$ we find

$$H_{\text{eff}} = -\frac{U_c}{4z}u^2 \sum_{\langle ij \rangle} \left(b_i^\dagger b_j + b_i b_j + \text{h.c.}\right) + \frac{U_c}{2}\sum_i b_i^\dagger b_i \qquad (A.3)$$

and for $u > 1$

$$H_{\text{eff}} = -\frac{U_c}{4z}\sum_{\langle ij \rangle} \left(b_i^\dagger b_j + b_i b_j + \text{h.c.}\right) + \frac{U}{2}\sum_i b_i^\dagger b_i. \qquad (A.4)$$

for chapter 3

## A.1.3 Bogoliubov transformation

In order to diagonalize the effective Hamiltonian $H_{\text{eff}}$ we first go to momentum space. We have ($z$ = number of nearest neighbors)

$$\sum_{\langle ij \rangle} (b_i^\dagger b_j + b_j^\dagger b_i) = -\frac{z}{2} \sum_k \left( b_\mathbf{k}^\dagger b_\mathbf{k} + b_{-\mathbf{k}} b_{-\mathbf{k}}^\dagger - 1 \right) \gamma_\mathbf{k}$$

$$\sum_{\langle ij \rangle} \left( b_i b_j + b_i^\dagger b_j^\dagger \right) = -\frac{z}{2} \sum_k \left( b_\mathbf{k} b_{-\mathbf{k}} + b_\mathbf{k}^\dagger b_{-\mathbf{k}}^\dagger \right) \gamma_\mathbf{k}$$

$$\sum_i b_i^\dagger b_i = \frac{1}{2} \sum_\mathbf{k} \left( b_\mathbf{k}^\dagger b_\mathbf{k} + b_{-\mathbf{k}} b_{-\mathbf{k}}^\dagger - 1 \right)$$

$$\gamma_\mathbf{k} = -\frac{2}{z} \sum_{i=1}^{z/2} \cos k_i$$

so that for $u \leq 1$

$$H_{\text{eff}} = \frac{U_c}{4} \sum_\mathbf{k} \mathcal{B}_\mathbf{k}^\dagger \begin{pmatrix} \frac{u^2}{2}\gamma_\mathbf{k} + 1 & \frac{u^2}{2}\gamma_\mathbf{k} \\ \frac{u^2}{2}\gamma_\mathbf{k} & \frac{u^2}{2}\gamma_\mathbf{k} + 1 \end{pmatrix} \mathcal{B}_\mathbf{k} - \frac{U_c}{4} N_s \quad (A.5)$$

and for $u > 1$

$$H_{\text{eff}} = \frac{U_c}{4} \sum_\mathbf{k} \mathcal{B}_\mathbf{k}^\dagger \begin{pmatrix} \frac{1}{2}\gamma_\mathbf{k} + u & \frac{1}{2}\gamma_\mathbf{k} \\ \frac{1}{2}\gamma_\mathbf{k} & \frac{1}{2}\gamma_\mathbf{k} + u \end{pmatrix} \mathcal{B}_\mathbf{k} - \frac{U}{4} N_s. \quad (A.6)$$

where

$$\mathcal{B}_\mathbf{k}^\dagger = \left( b_\mathbf{k}^\dagger, b_{-\mathbf{k}} \right) \quad \text{and} \quad \mathcal{B}_\mathbf{k} = \begin{pmatrix} b_\mathbf{k} \\ b_{-\mathbf{k}}^\dagger \end{pmatrix}. \quad (A.7)$$

The diagonalization of (A.5) and (A.7) is achieved by a Bogoliubov transformation

$$\begin{pmatrix} b_\mathbf{k} \\ b_{-\mathbf{k}}^\dagger \end{pmatrix} = \begin{pmatrix} \cosh \vartheta_\mathbf{k} & \sinh \vartheta_\mathbf{k} \\ \sinh \vartheta_\mathbf{k} & \cosh \vartheta_\mathbf{k} \end{pmatrix} \begin{pmatrix} \beta_\mathbf{k} \\ \beta_{-\mathbf{k}}^\dagger \end{pmatrix} \quad (A.8)$$

with

$$\vartheta_\mathbf{k} = \frac{1}{2} \text{atanh} \left( \frac{-u^2 \gamma_\mathbf{k}}{u^2 \gamma_\mathbf{k} + 2} \right) \quad \text{for} \quad u \leq 1 \quad (A.9)$$

and

$$\vartheta_\mathbf{k} = \frac{1}{2} \text{atanh} \left( \frac{-\gamma_\mathbf{k}}{\gamma_\mathbf{k} + 2u} \right) \quad \text{for} \quad u > 1 \quad (A.10)$$

*A.1 Pseudo-spin wave analysis*

which yields the pseudospin mode (3.54) as well as the renormalized ground-state energy (3.56).

### A.1.4  Matrix elements for the spectral density

Using the Lehmann representation (3.60) of the one-particle spectral density with the excited states obtained from the spin-wave analysis we arrive at the following expression for the one-particle spectral density

$$
\begin{aligned}
A_\sigma(\omega) &= \frac{1}{N_s} \sum_{\mathbf{q}} \Bigg\{ |\langle 0|2I_0^x|0\rangle|^2 \, \delta(\omega - g\varepsilon_{\mathbf{q}}) \\
&+ \sum_{\mathbf{k}} \Big[ (1 - n_{\mathbf{q}\sigma}) |\langle 0|2I_0^x|\mathbf{k}\rangle|^2 \, \delta(\omega - g\varepsilon_{\mathbf{q}} - \omega_{\mathbf{k}}) \\
&\quad + n_{\mathbf{q}\sigma} |\langle 0|2I_0^x|\mathbf{k}\rangle|^2 \, \delta(\omega - g\varepsilon_{\mathbf{q}} + \omega_{\mathbf{k}}) \Big] \\
&+ \sum_{\mathbf{k}\mathbf{k}'} \Big[ (1 - n_{\mathbf{q}\sigma}) |\langle 0|2I_0^x|\mathbf{k}\mathbf{k}'\rangle|^2 \, \delta(\omega - g\varepsilon_{\mathbf{q}} - \omega_{\mathbf{k}} - \omega_{\mathbf{k}'}) \\
&\quad + n_{\mathbf{q}\sigma} |\langle 0|2I_0^x|\mathbf{k}\mathbf{k}'\rangle|^2 \, \delta(\omega - g\varepsilon_{\mathbf{q}} + \omega_{\mathbf{k}} - \omega_{\mathbf{k}'}) \Big] \Bigg\}
\end{aligned}
$$

Here, $|\mathbf{k}\rangle$ denotes a pseudospin excitation with wave vector $\mathbf{k}$ and $|0\rangle$ is the vacuum for the spin-wave excitations. The first line is the contribution from the coherent quasiparticle band, $A_\sigma^{\text{coh}}(\omega)$, and the rest gives the contribution from incoherent excitations $A_\sigma^{\text{inc}}(\omega)$. The matrix elements are given by

$$
\begin{aligned}
\langle 0|2I_0^x|0\rangle &\equiv Z^{1/2} = \sqrt{\max(1 - u^2, 0)} \left[ 1 - \frac{1}{N_s} \sum_{\mathbf{k}} (\sinh \vartheta_{\mathbf{k}})^2 \right], \\
\langle 0|2I_0^x|\mathbf{k}\rangle &= -\min(u, 1) \frac{1}{\sqrt{N_s}} (\cosh \vartheta_{\mathbf{k}} + \sinh \vartheta_{\mathbf{k}}), \\
\langle 0|2I_0^x|\mathbf{k}\mathbf{k}'\rangle &= -2\sqrt{\max(1 - u^2, 0)} \frac{1}{N_s} (\cosh \vartheta_{\mathbf{k}} \sinh \vartheta_{\mathbf{k}'} + \cosh \vartheta_{\mathbf{k}'} \sinh \vartheta_{\mathbf{k}}).
\end{aligned}
$$

Numerical investigation shows that the contribution which involves two pseudospin excitations is small and in the following we will neglect it. We then find the coherent part (3.61) as well as the incoherent part (3.62) ($u \leq 1$) and (3.63) *for chapter 3*

($u > 1$). For this calculation we made use of the relation

$$\frac{d\omega_{\mathbf{k}}}{d\varepsilon_{\mathbf{k}}} = u^2 \frac{U_c}{4D} \frac{1}{\sqrt{1+u^2\varepsilon_{\mathbf{k}}/D}} = u^2 \frac{U_c}{4D} (\cosh\vartheta_{\mathbf{k}} + \sinh\vartheta_{\mathbf{k}})^2 \tag{A.11}$$

for $u \leq 1$ and a similar one for $u > 1$ to go to integration variables $y = \omega_{\mathbf{k}}$.

## A.1 Pseudo-spin wave analysis

# Appendix B

# for chapter 5

## B.1 Phenomenological thermoelectricity

As a supplementary information to Chap. 5 we discuss here aspects of the phenomenological thermoelectrical effects. We partly follow the lecture notes of Zlatić [149]. We limit ourselves to phenomena that are sufficiently close to equilibrium that the macroscopic intensive variables such as temperature $T(\vec{r})$, chemical potential $\mu(\vec{r})$ or electrostatic potential $\phi(\vec{r})$ are well defined locally. When gradients in these variables are maintained, currents of the corresponding density variables result. For a system of electrons we have from thermodynamics

$$du = Tds - e\bar{\phi}dn \tag{B.1}$$

where $u(\vec{r})$, $s(\vec{r})$ and $n(\vec{r})$ are the energy, entropy and particle density, respectively, and $\bar{\phi}(\vec{r}) = \phi(\vec{r}) - \mu(\vec{r})/e$ ($e > 0$ is the elementary charge). We write the phenomenological transport equations for the charge $\vec{J}_e$ and heat $\vec{J}_q$ currents in the presence of these gradient terms as

$$\vec{J}_e = \sigma\vec{\mathcal{E}} - \sigma S \vec{\nabla} T, \quad \vec{J}_q = ST\vec{J}_e - \kappa \vec{\nabla} T, \tag{B.2}$$

where the electrochemical field is $\vec{\mathcal{E}} = -\nabla(\phi + \mu/e) = -\nabla\bar{\phi}$. In (B.2) we have introduced the electrical conductivity $\sigma$, the thermal conductivity $\kappa$ and the Seebeck coefficient (thermopower) $S$. In general, these coefficients are tensors but for a cubic crystal we can consider them as scalars. The continuity equations

## B.1 Phenomenological thermoelectricity

for the charge and total energy current read

$$\dot{\rho} + \nabla \cdot \vec{J}_e = 0, \quad \dot{u} + \nabla \cdot \vec{J}_u = 0, \tag{B.3}$$

where $\vec{J}_u = \vec{J}_q + \bar{\phi}\vec{J}_e$. Eventually, in order to obtain a closed set of equations one has to relate the electrical potential to the charge and current densities by the Maxwell equations.

### B.1.1 Entropy production

The total change of the entropy in a volume $V$ is obtained by integrating (B.1) and using (B.2) and (B.3):

$$\frac{dS}{dt} = \int_V d^3r \frac{1}{T}\left(\dot{u} + e\bar{\phi}\dot{n}\right) = -\int_{\partial V} d\vec{\sigma} \cdot \left(\frac{\vec{J}_q}{T}\right) + \int_V d^3r \left[\frac{J_e^2}{\sigma T} + \frac{\kappa(\vec{\nabla}T)^2}{T^2}\right]. \tag{B.4}$$

Here, the first term gives the entropy change arising from an entropy current $\vec{J}_q/T$ flowing across the boundary $\partial V$ of $V$. The second term gives the volume contribution which is due to the change in the local entropy density. Thus, the entropy production in the volume $V$ is given by

$$\int_V d^3r \left[\frac{J_e^2}{\sigma T} + \frac{\kappa(\vec{\nabla}T)^2}{T^2}\right] \geq 0 \tag{B.5}$$

and is manifestly positive for $\sigma > 0$ and $\kappa > 0$. The first term describes the Joule heating and the second the energy dissipation due to the heat flow. Because thermoelectric effects are reversible, the Seebeck coefficient $S$ does not appear in (B.5): thermoelectric effects do not contribute to the overall entropy production.

### B.1.2 Temperature distribution

Using the transport equations (B.2) and the equations of continuity (B.3) for $u$ and $\rho$ we find for $\dot{\rho} = 0$:

$$\dot{u} = -\nabla \cdot \left(\vec{J}_q + \bar{\phi}\vec{J}_e\right) = \frac{J_e^2}{\sigma} + \nabla \cdot (\kappa \nabla T) - T\nabla S \cdot \vec{J}_e. \tag{B.6}$$

The first term describes the Joule heating, i.e. an electrical current flowing in an external field gives rise to the energy dissipation given by $J_e^2/\sigma$. The second

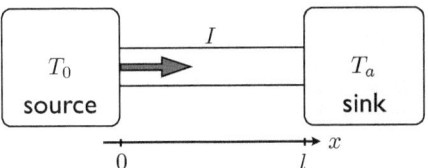

Figure B.1: Schematics of a thermoelectrical cooling device operating between two heat reservoirs. A battery drives a current $I$ between the source at a temperature $T_0$ and the sink at a temperature $T_a$.

term describes the energy dissipation due to the heat flow, $\nabla \cdot (\kappa \nabla T)$. The last term describes the production of the thermoelectric heat. Since it is linear in the current it can be either positive or negative and is thus reversible. Depending on the particular situation, the last term describes either the Thomson heat in an uniform sample $[\nabla S = (dS/dT)\nabla T]$ or the Peltier heat when the current crosses an interface $[\nabla S \sim (S_1 - S_2)]$. For $\dot{u} = 0$ we obtain the Domenicali equation [150]

$$0 = \frac{J_e^2}{\sigma} + \nabla \cdot (\kappa \nabla T) - T\nabla S \cdot \vec{J}_e \qquad (B.7)$$

which determines the temperature distribution in the stationary state when solved for given boundary conditions.

### B.1.3 Solution of transport equations

As an illustrative example we consider a "Gedankensetup" for a thermoelectric cooling device of length $l$ and cross section $A$ connecting two reservoirs at temperatures $T_0$ and $T_a$, see Fig. B.1 We assume that in the operating temperature range, the transport coefficients are constant. The goal is to find the coefficient of performance of such a device given by

$$\eta = \frac{J_q(x=0)A}{P} \qquad (B.8)$$

where $P$ is the power supplied by the battery to drive the current $I = J_e A$ and $\eta$ is optimized with respect to $I$. To find $P$ and the heat current at $x = 0$ we first

## B.1 Phenomenological thermoelectricity

have to calculate the temperature distribution which follows from Eq. (B.7):

$$T(y) = T_0 + (T_a - T_0)y + \frac{RI^2}{2K}y(1-y), \quad y = \frac{x}{l}, \tag{B.9}$$

where $R$ is the electrical resistance and $K$ the thermal conductance. Then, using (B.2) we find for the heat current at the source ($x = 0$) and at the sink ($x = l$)

$$AJ_q(x=0) = ST_0 I - K(T_a - T_0) - \frac{RI^2}{2}, \tag{B.10}$$

$$AJ_q(x=l) = ST_a I - K(T_a - T_0) + \frac{RI^2}{2}. \tag{B.11}$$

The first term describes the Peltier heat generated at the source and at the sink. The second term is the heat flux flowing from the hotter to the colder reservoir. The third term accounts for 1/2 of the total Joule heat generated in the total sample which is dumped into the reservoirs. In order to pump heat out of the source the Peltier heat has to overcome the irreversible heat production described by the second and third term. Certainly, for the cooling device, we have to choose $SI > 0$. In the stationary state, the difference between source and sink is supplied by the battery, thus

$$P = A\left[J_q(x=l) - J_q(x=0)\right] = S(T_a - T_0)I + RI^2. \tag{B.12}$$

The coefficient of performance for a fixed electrical current $I$ is

$$\eta(I) = \frac{J_q(x=0)A}{P} = \frac{ST_0 I - K(T_a - T_0) - \frac{RI^2}{2}}{S(T_a - T_0)I + RI^2}. \tag{B.13}$$

Maximizing with respect to $I$ yields

$$I_{opt} = \frac{K(T_a - T_0)}{ST_M}\left(1 + \sqrt{1 + Z_T T_M}\right) = \frac{S(T_a - T_0)}{R(\sqrt{1 + Z_T T_M} - 1)}. \tag{B.14}$$

and thus, the maximal coefficient of performance is

$$\eta_{max} = \eta(I_{opt}) = \frac{T_0}{T_a - T_0}\frac{\sqrt{1 + Z_T T_M} - \frac{T_a}{T_0}}{\sqrt{1 + Z_T T_M} + 1}. \tag{B.15}$$

Here, we have introduced the averaged temperature $T_M = (T_0 + T_a)/2$ and the thermoelectrical figure of merit

$$Z_T = \frac{S^2}{RK} = \frac{S^2\sigma}{\kappa}. \tag{B.16}$$

for chapter 5

For $Z_T T_M \to \infty$, $\eta_{max}$ reaches the Carnot efficiency. Thus, an efficient thermoelectric material should have a figure of merit $Z_T T_M$ as large as possible. This can be achieved by a high Seebeck coefficient $S$ while minimizing the loss due to Joule heating (small $R$) and irreversible heat conductance (small $K$). Typically, one can achieve at best $ZT_M \approx 1$ with real materials. Assuming $T_a/T_0 \approx 1.2$ one obtains about 9% of the Carnot efficiency.

## B.2 Kubo formalism for transport coefficients

We consider a single-band Hubbard model in the presence of a confining single-particle potential $V_i$

$$H = -t \sum_{\langle i,j \rangle, \sigma} c_{i\sigma}^\dagger c_{j\sigma} + U \sum_i n_{i\uparrow} n_{i\downarrow} + \sum_{i,\sigma} V_i n_{i\sigma}. \qquad (B.17)$$

The potential $V_i$ might be obtained from the self-consistent mean-field treatment of the electrostatic problem. In the following we will assume that the potential varies only along the $z$ direction and stays constant along $x$ and $y$ directions, thus $V_i \equiv V_l$ where $l$ is the layer index.

### B.2.1 Linear dynamical laws

We focus on the transport in the $x$-direction. Instead of using the form B.2 we write the linear dynamical laws in a symmetric way as

$$J_1^x = M^{11} \left[ \frac{e}{T} \frac{\partial \phi}{\partial x} - \frac{\partial}{\partial x}\left(\frac{\mu}{T}\right)\right] + M^{12} \frac{\partial}{\partial x}\left(\frac{1}{T}\right), \qquad (B.18)$$

$$J_2^x = M^{21} \left[ \frac{e}{T} \frac{\partial \phi}{\partial x} - \frac{\partial}{\partial x}\left(\frac{\mu}{T}\right)\right] + M^{22} \frac{\partial}{\partial x}\left(\frac{1}{T}\right). \qquad (B.19)$$

Here, $J_1^x$ and $J_2^x$ denote the x-component of the particle and the energy current, respectively. The symmetric coefficients $M^{uv}$ are related to the phenomenological

## B.2 Kubo formalism for transport coefficients

transport coefficients:

$$\sigma = \frac{e}{T}M^{11} \tag{B.20}$$

$$S = \frac{\mu}{eT} - \frac{M^{12}}{M^{11}}\frac{1}{Te} \tag{B.21}$$

$$\kappa^e = \frac{1}{T^2}\left[M^{22} - \frac{(M^{12})^2}{M^{11}}\right]. \tag{B.22}$$

In the linear response Kubo formalism the coefficients $M^{uv}$ are given by

$$M^{uv} = T\int_0^\infty dt \int_0^\beta d\tau \mathrm{Tr}\left[\rho_0 j_v^x(-t-i\tau)j_u^x(0)\right] \tag{B.23}$$

where $j_1^x$ and $j_2^x$ are the particle and heat current operators in $x$ direction, respectively. For the model (B.17) they are given by the following operators

$$\mathbf{j}_1 = \sum_{\mathbf{k},l,\sigma}\mathbf{v_k}c^\dagger_{\mathbf{k}l\sigma}c_{\mathbf{k}l\sigma} \tag{B.24}$$

$$\mathbf{j}_2 = \sum_{\mathbf{k},l,\sigma}(\varepsilon_\mathbf{k}+V_l)\mathbf{v_k}c^\dagger_{\mathbf{k}l\sigma}c_{\mathbf{k}l\sigma}$$

$$+\frac{U}{2}\sum_{\mathbf{k}\mathbf{q}\mathbf{q}',l}[\mathbf{v_q}+\mathbf{v_{q'}}]c^\dagger_{\mathbf{q}l\uparrow}c_{\mathbf{q}'l\uparrow}c^\dagger_{\mathbf{k}l\downarrow}c_{\mathbf{k}+\mathbf{q}-\mathbf{q}'l\downarrow} \tag{B.25}$$

where $\mathbf{v_k} = \nabla\varepsilon_\mathbf{k}$ ($\hbar = 1$), $\mathbf{k}$ is a two-dimensional crystal momentum and $l$ the layer index.

### B.2.2 Quasiparticle transport

In order to clarify the approximations undertaken in the Boltzmann description for the dc transport coefficients, we derive the transport distribution function $\Phi(E)$ starting from the linear-response Kubo formula (B.23) and assuming the validity of the quasiparticle description outlined in Sec. 4.2. It is convenient to evaluate the current-current correlation function at finite temperature and for a finite frequency [142]

$$\pi(i\omega_n) = -\frac{e^2}{2V}\int_0^\beta e^{i\omega_n\tau}\langle T_\tau\mathbf{j}_1(\tau)\cdot\mathbf{j}_1(0)\rangle, \tag{B.26}$$

for chapter 5

where $V$ is the volume of the sample. A subsequent analytical continuation $i\omega_n \to \omega + i\delta$ yields the dc conductivity by taking the following limit

$$\sigma = -\lim_{\omega \to 0}\left[\frac{\operatorname{Im}\pi^R(\omega)}{\omega}\right]. \qquad (B.27)$$

The remaining transport coefficients may be obtained by applying the Jonson-Mahan theorem [151] which holds for the model (B.17) [152]. In the following we show that when neglecting vertex corrections in Eq. (B.26), one arrives at the transport distribution function [Eq. (5.21)]. Thus, we approximate

$$\pi(i\omega_n) = \frac{e^2}{V\beta}\sum_{\mathbf{k}}|\nabla\varepsilon_{\mathbf{k}}|^2 \sum_{ip_m}\operatorname{Tr}\left[\hat{\mathcal{G}}(\mathbf{k},ip_m+i\omega_n)\hat{\mathcal{G}}(\mathbf{k},ip_m)\right],$$

where the trace has to be performed over the different layers and $\mathcal{G}$ is the full Matsubara Green's function including all self-energy corrections. After letting $i\omega_n \to \omega + i\delta$, one finds the form given in Eq. (5.24),

$$\sigma = \int dE \left(-\frac{\partial f}{\partial E}\right)\Phi(E) \qquad (B.28)$$

with the transport distribution function

$$\Phi(E) = \frac{\pi e^2}{V}\sum_{\mathbf{k}}|\nabla\varepsilon_{\mathbf{k}}|^2 \operatorname{Tr}\left[\hat{A}(\mathbf{k},E)^2\right]. \qquad (B.29)$$

Here, we have introduced the spectral density matrix $\hat{A}(\mathbf{k},E) = -\frac{1}{\pi}\operatorname{Im}\hat{G}^R(\mathbf{k},E)$ for which we assume (see Sec. 4.2) the following low-energy form

$$\left[\hat{A}(\mathbf{k},E)\right]_{ll'} = \frac{1}{\pi}\sum_{\nu}\frac{z_l\psi_{\mathbf{k}\nu}(l)\gamma_{\mathbf{k}\nu}\psi_{\mathbf{k}\nu}(l')z_{l'}}{(E-E_{\mathbf{k}\nu})^2+\gamma_{\mathbf{k}\nu}^2}. \qquad (B.30)$$

The quasiparticle life-time is given by the relation $\gamma_{\mathbf{k}\nu} = 1/(2\tau_{\mathbf{k}\nu})$, where

$$\gamma_{\mathbf{k}\nu} = -\sum_{l}z_l^2\psi_{\mathbf{k}\nu}(l)^2\left[\operatorname{Im}\hat{\Sigma}(E_{\mathbf{k}\nu})\right]_{ll}. \qquad (B.31)$$

In lowest order in temperature we can restrict the analysis to the Fermi surface. In this case $\gamma_{\mathbf{k}\nu}$ is proportional to the impurity concentration $c_{\text{imp}}$ for dilute impurities. For non-degenerate subbands and in the limit $c_{\text{imp}} \to 0$ only terms,

## B.2 Kubo formalism for transport coefficients

which are diagonal in the subband index, contribute to the trace in Eq. (B.29). After using the relation (4.45) for the renormalization amplitude $Z_{\mathbf{k}\nu}$ one arrives at Eq. (5.21) obtained from the linearized Boltzmann equation. In actual calculations we estimate $\gamma_{\mathbf{k}\nu}$ by the approximations discussed in Sec. 5.3.2.

In summary, we have assumed the validity of a Fermi-liquid description with local self-energy corrections, dominant $s$-wave scattering on dilute impurities, and that effects of weak localization can be neglected. Under these conditions we expect that, referring to the homogenous and isotropic system [153] the transport distribution function used in the Sec. 5.3.1 captures the main features in lowest order in temperature.

### B.2.3 Approach from the atomic limit

In the following we will work in the atomic limit $t \to 0$. This is a reasonable starting point in the high-temperature limit where the kinetic energy can be neglected compared to temperature and we are dealing effectively with a classical lattice gas. Furthermore, we will work in the strongly correlated regime defined as $t \ll k_B T \ll U$. In the atomic limit, the relation (2.20) between the chemical potential and the particle density holds for each layer separately

$$n_l = \frac{2\xi_l + 2\xi_l^2 e^{-\beta U}}{1 + 2\xi_l + \xi_l^2 e^{-\beta U}}. \tag{B.32}$$

where $\xi_l = e^{\beta(\mu - V_l)}$. In the limit $U \to \infty$ we find

$$\mu - V_l = k_B T \log \xi_l = -k_B T \log \left[ \frac{2(1 - n_l)}{n_l} \right]. \tag{B.33}$$

This relation will be important in order to determine the total thermopower. In lowest order in the hopping amplitude $t$ and suppressing the contribution from doubly occupied sites by letting $U \to \infty$ there is the following simple relation between the particle and energy current operators in layer $l$

$$\mathbf{j}_{2,l} = V_l \mathbf{j}_{1,l}. \tag{B.34}$$

Here, $\mathbf{j}_{u,l}$ is defined by $\mathbf{j}_u = \sum_l \mathbf{j}_{u,l}$. In this way we find the relation

$$M^{12}(l) = V_l M^{11}(l) \tag{B.35}$$

and the thermopower can be written as

$$S = \sum_l \frac{\mu - V_l}{eT} \frac{\sigma_l}{\sigma} \equiv \sum_l \frac{S_l \sigma_l}{\sigma}. \tag{B.36}$$

As in the homogeneous case $S_l$ can be expressed in terms of the particle density in layer $l$

$$S_l = \frac{\mu - V_l}{eT} = \frac{k_B}{e} \log \xi_l = -\frac{k_B}{e} \log \left[ \frac{2(1-n_l)}{n_l} \right]. \tag{B.37}$$

The remaining task is thus the calculation of the ratio $\sigma_l/\sigma$ where $\sigma_l = eM^{11}(l)/T$. We note here that the conductivity is infinite in the atomic limit since relaxation processes are missing. However, the ratio $\sigma_l/\sigma$ is finite. In fact, we can use the expression given by Mukerjee [147] to express the conductivity of layer $l$ as:

$$\sigma_l = \frac{e^2 A \beta}{2} n_l (1 - n_l) \tag{B.38}$$

with a temperature and doping independent constant $A$ and we find

$$\frac{\sigma_l}{\sigma} = \frac{n_l(1-n_l)}{\sum_l n_l(1-n_l)}. \tag{B.39}$$

This yields the result

$$S = -\frac{k_B}{e} \frac{\sum_l \log\left[\frac{2(1-n_l)}{n_l}\right] n_l(1-n_l)}{\sum_l n_l(1-n_l)}. \tag{B.40}$$

In this expression $n_l$ has to be found numerically for a given potential. Equation (B.40) corresponds to the Heikes limit generalized to inhomogeneous systems. In this limit, the thermopower depends only on the spatial configuration of the charge. In a similar way, the powerfactor may also be considered as a functional of the charge distribution

$$PF = Q^2 \sigma = \frac{Ak_B}{2T} \frac{\left[\sum_l \log\left[\frac{2(1-n_l)}{n_l}\right] n_l(1-n_l)\right]^2}{(N+M) \sum_l n_l(1-n_l)} \tag{B.41}$$

where for the above expression we have assumed a superlattice with parameters $N$ and $M$.

## B.2 Kubo formalism for transport coefficients

# Bibliography

[1] D. Pines and P. Nozières, *The Theory of Quantum Liquids* (Perseus Books Publishing, 1966).

[2] L. D. Landau, *Theory of the Fermi-liquid*, Sov. Phys. JETP **3**, 920 (1957).

[3] P. W. Anderson, *Basic notions of condensed matter physics* (Perseus Books Publishing, 1997).

[4] P. Hohenberg and W. Kohn, *Inhomogeneous Electron Gas*, Phys. Rev. **136**, B864 (1964).

[5] W. Kohn and L. J. Sham, *Self-Consistent Equations Including Exchange and Correlation Effects*, Phys. Rev. **140**, A1133 (1965).

[6] N. F. Mott, *The Basis of the Electron Theory of Metals, with Special Reference to the Transition Metals*, Proc. Roy. Soc. A **62**, 416 (1949).

[7] J. Hubbard, *Electron Correlations in Narrow Energy Bands*, Proc. Roy. Soc. A **276**, 238 (1963).

[8] J. G. Bednorz and K. A. Müller, *Possible high-$T_c$ superconductivity in the BaLaCuO system*, Z. Phys. B **64**, 189 (1986).

[9] D. C. Tsui, H. L. Stormer, and A. C. Gossard, *Two-Dimensional Magnetotransport in the Extreme Quantum Limit*, Phys. Rev. Lett. **48**, 1559 (1982).

[10] R. B. Laughlin, *Anomalous Quantum Hall Effect: An Incompressible Quantum Fluid with Fractionally Charged Excitations*, Phys. Rev. Lett. **50**, 1395 (1983).

## BIBLIOGRAPHY

[11] M. Imada, A. Fujimori, and Y. Tokura, *Metal-insulator transitions*, Rev. Mod. Phys. **70**, 1039 (1998).

[12] E. Dagotto, *Complexity in Strongly Correlated Electronic Systems*, Science **309**, 257 (2005).

[13] A. Ohtomo, D. A. Muller, J. L. Grazul, and H. Y. Hwang, *Artificial charge-modulation in atomic-scale perovskite titanate superlattices*, Nature **419**, 378 (2002).

[14] A. Ohtomo and H. Y. Hwang, *A high-mobility electron gas at the $LaAlO_3/SrTiO_3$ heterointerface*, Nature **427**, 423 (2004).

[15] S. Thiel, G. Hammerl, A. Schmehl, W. Schneider, and J. Mannhart, *Tunable Quasi-Two-Dimensional Electron Gases in Oxide Heterostructures*, Science **313**, 1942 (2006).

[16] N. Reyren, S. Thiel, A. D. Caviglia, L. F. Kourkoutis, G. Hammerl, C. Richter, C. W. Schneider, T. Kopp, A.-S. Ruetschi, D. Jaccard, et al., *Superconducting Interfaces Between Insulating Oxides*, Science **317**, 1196 (2007).

[17] A. D. Caviglia, S. Gariglio, N. Reyren, D. Jaccard, T. Schneider, M. Gabay, S. Thiel, G. Hammerl, J. Mannhart, and J. M. Triscone, *Electric field control of the $LaAlO_3/SrTiO_3$ interface ground state*, Nature **456**, 624 (2008).

[18] W. C. Sheets, B. Mercey, and W. Prellier, *Effect of charge modulation in $(LaVO_3)_m(SrVO_3)_n$ superlattices on the insulator-metal transition*, Appl. Phys. Lett. **91**, 192102 (2007).

[19] Y. Hotta, T. Susaki, and H. Y. Hwang, *Polar Discontinuity Doping of the $LaVO_3/SrTiO_3$ Interface*, Phys. Rev. Lett. **99**, 236805 (2007).

[20] T. Higuchi, Y. Hotta, T. Susaki, A. Fujimori, and H. Y. Hwang, *Modulation doping of a Mott quantum well by a proximate polar discontinuity*, Phys. Rev. B **79**, 075415 (2009).

[21] H. Ibach, *Physics of surfaces and interfaces* (Springer-Verlag Berlin Heidelberg New York, 2006).

# BIBLIOGRAPHY

[22] M. Potthoff and W. Nolting, *Metallic surface of a Mott insulator – Mott insulating surface of a metal*, Phys. Rev. B **60**, 7834 (1999).

[23] N. Nakagawa, H. Y. Hwang, and D. A. Muller, *Why some interfaces cannot be sharp*, Nat. Mater. **5**, 204 (2006).

[24] P. A. Salvador, A.-M. Haghiri-Gosnet, B. Mercey, M. Hervieu, and B. Raveau, *Growth and magnetoresistive properties of $(LaMnO_3)_m(SrMnO_3)_n$ superlattices*, Appl. Phys. Lett. **75**, 2638 (1999).

[25] A. Bhattacharya, S. J. May, S. G. E. te Velthuis, M. Warusawithana, X. Zhai, B. Jiang, J.-M. Zuo, M. R. Fitzsimmons, S. D. Bader, and J. N. Eckstein, *Metal-Insulator Transition and Its Relation to Magnetic Structure in $(LaMnO_3)_{2n}/(SrMnO_3)_n$ Superlattices*, Phys. Rev. Lett. **100**, 257203 (2008).

[26] C. Noguera, *Polar oxide surfaces*, J. Phys. Cond. Mat. **12**, R367 (2000).

[27] R. Hesper, L. H. Tjeng, A. Heeres, and G. A. Sawatzky, *Photoemission evidence of electronic stabilization of polar surfaces in $K_3C_{60}$*, Phys. Rev. B **62**, 16046 (2000).

[28] S. Okamoto and A. J. Millis, *Electronic reconstruction at an interface between a Mott insulator and a band insulator*, Nature **428**, 630 (2004).

[29] S. A. Pauli and P. R. Willmott, *Conducting interfaces between polar and non-polar insulating perovskites*, J. Phys. Cond. Mat. **20**, 264012 (2008).

[30] S. Okamoto, A. J. Millis, and N. A. Spaldin, *Lattice Relaxation in Oxide Heterostructures: $LaTiO_3/SrTiO_3$ Superlattices*, Phys. Rev. Lett. **97**, 056802 (2006).

[31] D. R. Hamann, D. A. Muller, and H. Y. Hwang, *Lattice-polarization effects on electron-gas charge densities in ionic superlattices*, Phys. Rev. B **73**, 195403 (2006).

[32] P. Larson, Z. S. Popović, and S. Satpathy, *Lattice relaxation effects on the interface electron states in the perovskite oxide heterostructures: $LaTiO_3$ monolayer embedded in $SrTiO_3$*, Phys. Rev. B **77**, 245122 (2008).

# BIBLIOGRAPHY

[33] A. Brinkman, M. Huijben, M. van Zalk, J. Huijben, U. Zeitler, J. C. Maan, W. G. van der Wiel, G. Rijnders, D. H. A. Blank, and H. Hilgenkamp, *Magnetic effects at the interface between non-magnetic oxides*, Nat. Mater. **6**, 493 (2007).

[34] J. Chakhalian, J. W. Freeland, H.-U. Habermeier, G. Cristiani, G. Khaliullin, M. van Veenendaal, and B. Keimer, *Orbital Reconstruction and Covalent Bonding at an Oxide Interface*, Science **318**, 1114 (2007).

[35] Z. S. Popovic and S. Satpathy, *Wedge-Shaped Potential and Airy-Function Electron Localization in Oxide Superlattices*, Phys. Rev. Lett. **94**, 176805 (2005).

[36] Z. S. Popovic, S. Satpathy, and R. M. Martin, *Origin of the Two-Dimensional Electron Gas Carrier Density at the $LaAlO_3$ on $SrTiO_3$ Interface*, Phys. Rev. Lett. **101**, 256801 (2008).

[37] W.-C. Lee and A. H. MacDonald, *Modulation doping near Mott-insulator heterojunctions*, Phys. Rev. B **74**, 075106 (2006).

[38] W.-C. Lee and A. H. MacDonald, *Electronic interface reconstruction at polar-nonpolar Mott-insulator heterojunctions*, Phys. Rev. B **76**, 075339 (2007).

[39] S. Okamoto and A. J. Millis, *Spatial inhomogeneity and strong correlation physics: A dynamical mean-field study of a model Mott-insulator–band-insulator heterostructure*, Phys. Rev. B **70**, 241104(R) (2004).

[40] S. Okamoto and A. J. Millis, *Theory of Mott insulator–band insulator heterostructures*, Phys. Rev. B **70**, 075101 (2004).

[41] S. Okamoto and A. J. Millis, *Interface ordering and phase competition in a model Mott-insulator-band-insulator heterostructure*, Phys. Rev. B **72**, 235108 (2005).

[42] S. S. Kancharla and E. Dagotto, *Metallic interface at the boundary between band and Mott insulators*, Phys. Rev. B **74**, 195427 (2006).

## BIBLIOGRAPHY

[43] J. K. Freericks, *Transport in multilayered nanostructures: the dynamical mean-field approach* (Imperial College Press, 2006).

[44] N. Pavlenko and T. Kopp, *Interface Controlled Electronic Charge Inhomogeneities in Correlated Heterostructures*, Phys. Rev. Lett. **97**, 187001 (2006).

[45] N. Pavlenko and T. Kopp, *Electrostatic interface tuning in correlated superconducting heterostructures*, Phys. Rev. B **72**, 174516 (2005).

[46] G. Kotliar, S. Y. Savrasov, K. Haule, V. S. Oudovenko, O. Parcollet, and C. A. Marianetti, *Electronic structure calculations with dynamical mean-field theory*, Rev. Mod. Phys. **78**, 865 (2006).

[47] H. Ishida and A. Liebsch, *Origin of metallicity of $LaTiO_3/SrTiO_3$ heterostructures*, Phys. Rev. B **77**, 115350 (2008).

[48] S. Okamoto and A. J. Millis, in *Proceedings of SPIE Conference on Strongly Correlated Electron Materials: Physics and Nanoengineering* (2005).

[49] G. Beni, *Thermoelectric power of the narrow-band Hubbard chain at arbitrary electron density: Atomic limit*, Phys. Rev. B **10** (1974).

[50] R. Frésard and P. Wölfle, *Unified Slave Boson Representation of Spin and Charge Degrees of Freedom for Strongly Correlated Fermi Systems*, Int. J. Mod. Phys. B **6**, 237 (1992).

[51] T. Holstein and H. Primakoff, *Field Dependence of the Intrinsic Domain Magnetization of a Ferromagnet*, Phys. Rev. **58** (1940).

[52] A. Auerbach, *Interacting electrons and quantum magnetism* (Springer-Verlag New York, 1994).

[53] C. Jayaprakash, H. R. Krishnamurthy, and S. Sarker, *Mean-field theory for the t-J model*, Phys. Rev. B **40**, 2610 (1989).

[54] A. Tsevlik, *Quantum field theory in condensed matter physics* (Cambridge University Press, 2003).

# BIBLIOGRAPHY

[55] D. Yoshioka, *Slave-Fermion Mean Field Theory of the Hubbard Model*, J. Phys. Soc. Jpn. **58**, 1516 (1989).

[56] S. E. Barnes, *New method for the Anderson model*, J. Phys. F **6**, 1375 (1976).

[57] P. Coleman, *New approach to the mixed-valence problem*, Phys. Rev. B **29**, 3035 (1984).

[58] P. Coleman, *Mixed valence as an almost broken symmetry*, Phys. Rev. B **35**, 5072 (1987).

[59] N. Read and D. Newns, *On the solution of he Coqblin-Schrieffer Hamiltonian by the large-N expansion technique*, J. Phys. C **16**, 3473 (1983).

[60] N. Read and D. M. Newns, *A new functional integral formalism for the degenerate Anderson model*, J. Phys. C **16**, L1055 (1983).

[61] G. Kotliar and J. Liu, *Superexchange mechanism and d-wave superconductivity*, Phys. Rev. B **38**, 5142 (1988).

[62] X.-G. Wen and P. A. Lee, *Theory of Underdoped Cuprates*, Phys. Rev. Lett. **76**, 503 (1996).

[63] P. A. Lee, N. Nagaosa, and X.-G. Wen, *Doping a Mott insulator: Physics of high-temperature superconductivity*, Rev. Mod. Phys. **78**, 17 (2006).

[64] G. Kotliar and A. E. Ruckenstein, *New Functional Integral Approach to Strongly Correlated Fermi Systems: The Gutzwiller Approximation as a Saddle Point*, Phys. Rev. Lett. **57**, 1362 (1986).

[65] M. C. Gutzwiller, *Effect of Correlation on the Ferromagnetism of Transition Metals*, Phys. Rev. Lett. **10**, 159 (1963).

[66] W. F. Brinkman and T. M. Rice, *Application of Gutzwiller's Variational Method to the Metal-Insulator Transition*, Phys. Rev. B **2**, 4302 (1970).

[67] T. Li, P. Wölfle, and P. J. Hirschfeld, *Spin-rotation-invariant slave-boson approach to the Hubbard model*, Phys. Rev. B **40**, 6817 (1989).

# BIBLIOGRAPHY

[68] F. Lechermann, A. Georges, G. Kotliar, and O. Parcollet, *Rotationally invariant slave-boson formalism and momentum dependence of the quasiparticle weight*, Phys. Rev. B **76**, 155102 (2007).

[69] A. Rüegg, S. Pilgram, and M. Sigrist, *Strongly renormalized quasi-two-dimensional electron gas in a heterostructure with correlation effects*, Phys. Rev. B **75**, 195117 (2007).

[70] A. Rüegg, S. Pilgram, and M. Sigrist, *Aspects of metallic low-temperature transport in Mott-insulator/band-insulator superlattices: Optical conductivity and thermoelectricity*, Phys. Rev. B **77**, 245118 (2008).

[71] A. Rüegg and M. Sigrist, *Role of multiple subband renormalization in the electronic transport of correlated oxide superlattices* (2008), to appear as proceeding of the Hvar 2008 workshop.

[72] E. Arrigoni and G. C. Strinati, *Beyond the Gutzwiller approximation in the slave-boson approach: Inclusion of fluctuations with the correct continuum limit of the functional integral*, Phys. Rev. Lett. **71**, 3178 (1993).

[73] T. Jolicoeur and J. C. Le Guillou, *Fluctuations beyond the Gutzwiller approximation in the slave-boson approach*, Phys. Rev. B **44**, 2403 (1991).

[74] E. Arrigoni and G. C. Strinati, *Correct continuum limit of the functional-integral representation for the four-slave-boson approach to the Hubbard model: Paramagnetic phase*, Phys. Rev. B **52** (1995).

[75] R. Frésard and T. Kopp, *Exact results in a slave boson saddle point approach for a strongly correlated electron model*, Phys. Rev. B **78**, 4 (2008).

[76] G. Seibold, E. Sigmund, and V. Hizhnyakov, *Unrestricted slave-boson mean-field approximation for the two-dimensional Hubbard model*, Phys. Rev. B **57**, 6937 (1998).

[77] R. Frésard and P. Wölfle, *Spiral magnetic states in the large-U Hubbard model: a slave boson approach*, J. Phys. Cond. Mat. **4**, 3625 (1992).

[78] P. J. H. Denteneer and M. Blaauboer, *Helicity modulus and effective hopping in the two-dimensional Hubbard model using slave-boson methods*, J. Phys. Cond. Mat. **7**, 151 (1995).

[79] M. Raczkowski, R. Fresard, and A. M. Oles, *Slave-boson approach to the metallic stripe phases with large unit cells*, Phys. Rev. B **73**, 174525 (2006).

[80] A. Georges, G. Kotliar, W. Krauth, and M. J. Rozenberg, *Dynamical mean-field theory of strongly correlated fermion systems and the limit of infinite dimensions*, Rev. Mod. Phys. **68**, 13 (1996).

[81] P. W. Anderson, P. A. Lee, M. Randeria, T. M. Rice, N. Trivedi, and F. C. Zhang, *The physics behind high-temperature superconducting cuprates: the 'plain vanilla' version of RVB*, J. Phys. Cond. Mat. **16**, R755 (2004).

[82] B. Edegger, V. N. Muthukumar, and C. Gros, *Gutzwiller–RVB theory of high-temperature superconductivity: Results from renormalized mean-field theory and variational Monte Carlo calculations*, Adv. Phys. **56**, 927 (2007).

[83] D. Vollhardt, *Normal $^3$He: an almost localized Fermi liquid*, Rev. Mod. Phys. **56**, 99 (1984).

[84] D. Baeriswyl, C. Gros, and T. M. Rice, *Landau parameters of almost-localized Fermi liquids*, Phys. Rev. B **35**, 8391 (1987).

[85] P. W. Anderson, *The Resonating Valence Bond State in $La_2CuO_4$ and Superconductivity*, Science **235**, 1196 (1987).

[86] F. Gebhard, *Gutzwiller correlated wave functions in finite dimensions d: A systematic expansion in 1/d*, Phys. Rev. B **41**, 9452 (1990).

[87] F. C. Zhang, C. Gros, T. M. Rice, and H. Shiba, *A renormalised Hamiltonian approach to a resonant valence bond wavefunction*, Superconductor Science and Technology **1**, 36 (1988).

[88] A. Rüegg, M. Indergand, S. Pilgram, and M. Sigrist, *Slave-boson mean-field theory of the Mott transition in the two-band Hubbard model*, Eur. Phys. J. B **48**, 55 (2005).

[89] A. Koga, N. Kawakami, T. M. Rice, and M. Sigrist, *Orbital-Selective Mott Transitions in the Degenerate Hubbard Model*, Phys. Rev. Lett. **92**, 216402 (2004).

[90] L. de'Medici, A. Georges, and S. Biermann, *Orbital-selective Mott transition in multiband systems: Slave-spin representation and dynamical mean-field theory*, Phys. Rev. B **72**, 205124 (2005).

[91] A. Koga, N. Kawakami, T. M. Rice, and M. Sigrist, *Spin, charge, and orbital fluctuations in a multiorbital Mott insulator*, Phys. Rev. B **72**, 045128 (2005).

[92] A. Liebsch, *Mott Transitions in Multiorbital Systems*, Phys. Rev. Lett. **91**, 226401 (2003).

[93] A. Liebsch, *Single Mott transition in the multiorbital Hubbard model*, Phys. Rev. B **70**, 165103 (2004).

[94] V. I. Anisimov, I. A. Nekrasov, D. E. Kondakov, T. M. Rice, and M. Sigrist, *Orbital-selective Mott-insulator transition in $Ca_{2-x}Sr_xRuO_4$*, Eur. Phys. J. B **25**, 191 (2002).

[95] M. Sigrist and M. Troyer, *Orbital and spin correlations in $Ca_{2-x}Sr_xRuO_4$: A mean field study*, Eur. Phys. J. B **39**, 207 (2004).

[96] R. Roldan, A. Rüegg, and M. Sigrist, *Interplay of metamagnetic and structural transitions in $Ca_{2-x}Sr_xRuO_4$*, Eur. Phys. J. B **64**, 185 (2008).

[97] S. D. Huber and A. Rüegg, *Dynamically Generated Double Occupancy as a Probe of Cold Atom Systems*, Phys. Rev. Lett. **102**, 065301 (2009).

[98] C. Castellani, G. Kotliar, R. Raimondi, M. Grilli, Z. Wang, and M. Rozenberg, *Collective excitations, photoemission spectra, and optical gaps in strongly correlated Fermi systems*, Phys. Rev. Lett. **69**, 2009 (1992).

[99] R. Raimondi and C. Castellani, *Lower and upper Hubbard bands: A slave-boson treatment*, Phys. Rev. B **48**, 11453 (1993).

# BIBLIOGRAPHY

[100] M. Lavagna, *Functional-integral approach to strongly correlated Fermi systems: Quantum fluctuations beyond the Gutzwiller approximation*, Phys. Rev. B **41**, 142 (1990).

[101] S. Florens and A. Georges, *Slave-rotor mean-field theories of strongly correlated systems and the Mott transition in finite dimensions*, Phys. Rev. B **70**, 035114 (2004).

[102] W. F. Brinkman and T. M. Rice, *Single-Particle Excitations in Magnetic Insulators*, Phys. Rev. B **2**, 1324 (1970).

[103] D. Tanasković, V. Dobrosavljević, E. Abrahams, and G. Kotliar, *Disorder Screening in Strongly Correlated Systems*, Phys. Rev. Lett. **91**, 066603 (2003).

[104] M. C. O. Aguiar, V. Dobrosavljević, E. Abrahams, and G. Kotliar, *Scaling behavior of an Anderson impurity close to the Mott-Anderson transition*, Phys. Rev. B **73**, 115117 (2006).

[105] P. Werner, A. Comanac, L. de' Medici, M. Troyer, and A. J. Millis, *Continuous-Time Solver for Quantum Impurity Models*, Phys. Rev. Lett. **97**, 076405 (2006).

[106] R. Bulla, T. A. Costi, and T. Pruschke, *Numerical renormalization group method for quantum impurity systems*, Rev. Mod. Phys. **80**, 395 (2008).

[107] S. Y. Savrasov, V. Oudovenko, K. Haule, D. Villani, and G. Kotliar, *Interpolative approach for solving the Anderson impurity model*, Phys. Rev. B **71**, 115117 (2005).

[108] S. Florens and A. Georges, *Quantum impurity solvers using a slave rotor representation*, Phys. Rev. B **66**, 165111 (2002).

[109] Z. J. Ning, L. Wang, Z. Fang, and X. Dai, *A Fast Impurity Solver Based on Gutzwiller variational approach* (2008), arXiv:0810.2385.

[110] M. Takizawa, H. Wadati, K. Tanaka, M. Hashimoto, T. Yoshida, A. Fujimori, A. Chikamatsu, H. Kumigashira, M. Oshima, K. Shibuya, et al., *Photoemission from Buried Interfaces in $SrTiO_3/LaTiO_3$ Superlattices*, Phys. Rev. Lett. **97**, 057601 (2006).

[111] S. S. A. Seo, W. S. Choi, H. N. Lee, L. Yu, K. W. Kim, C. Bernhard, and T. W. Noh, *Optical Study of the Free-Carrier Response of $LaTiO_3/SrTiO_3$ Superlattices*, Phys. Rev. Lett. **99**, 266801 (2007).

[112] S. Schwieger, M. Potthoff, and W. Nolting, *Correlation and surface effects in vanadium oxides*, Phys. Rev. B **67**, 165408 (2003).

[113] S. Wehrli, D. Poilblanc, and T. M. Rice, *Charge profile of surface doped $C_{60}$*, Eur. Phys. J. B **23**, 345 (2001).

[114] B. Moller, K. Doll, and R. Frésard, *A slave-boson approach to ferromagnetism in the large-U Hubbard model*, J. Phys. Cond. Mat. **5**, 4847 (1993).

[115] H. Tsunetsugu, M. Sigrist, and K. Ueda, *The ground-state phase diagram of the one-dimensional Kondo lattice model*, Rev. Mod. Phys. **69**, 809 (1997).

[116] T. M. Rice, K. Ueda, H. R. Ott, and H. Rudigier, *Normal-state properties of heavy-electron systems*, Phys. Rev. B **31** (1985).

[117] T. M. Rice and K. Ueda, *Gutzwiller Variational Approximation to the Heavy-Fermion Ground State of the Periodic Anderson Model*, Phys. Rev. Lett. **55**, 995 (1985).

[118] A. C. Hewson, *The Kondo problem to heavy fermions* (Cambridge University Press, 1993).

[119] J. H. Van Vleck, *Note on the Interactions between the Spins of Magnetic Ions or Nuclei in Metals*, Rev. Mod. Phys. **34**, 681 (1962).

[120] M. Potthoff, *Two-site dynamical mean-field theory*, Phys. Rev. B **64**, 12 (2001).

[121] M. Potthoff, *Self-energy-functional approach: Analytic results and the Mott-Hubbard transition*, Eur. Phys. J. B **36**, 335 (2003).

## BIBLIOGRAPHY

[122] J. Bunemann and F. Gebhard, *Equivalence of Gutzwiller and slave-boson mean-field theories for multiband Hubbard models*, Phys. Rev. B **76**, 193104 (2007).

[123] W.-H. Ko, C. P. Nave, and P. A. Lee, *Extended Gutzwiller approximation for inhomogeneous systems*, Phys. Rev. B **76**, 245113 (2007).

[124] S. Pilgram, *Gutzwiller approximation for the Hubbard heterostructure* (2008), personal communication.

[125] Y. Tokura, Y. Taguchi, Y. Okada, Y. Fujishima, T. Arima, K. Kumagai, and Y. Iye, *Filling dependence of electronic properties on the verge of metal–Mott-insulator transition in $Sr_{1-x}La_xTiO_3$*, Phys. Rev. Lett. **70**, 2126 (1993).

[126] K. Shibuya, T. Ohnishi, M. Kawasaki, H. Koinuma, and M. Lippmaa, *Metallic $LaTiO_3/SrTiO_3$ Superlattice Films on the $SrTiO_3$ (100) Surface*, Jpn. J. Appl. Phys. **43**, L1178 (2004).

[127] T. Okuda, K. Nakanishi, S. Miyasaka, and Y. Tokura, *Large thermoelectric response of metallic perovskites: $Sr_{1-x}La_xTiO_3$ ($0 \lesssim x \lesssim 0.1$)*, Phys. Rev. B **63**, 113104 (2001).

[128] P. M. Chaikin and G. Beni, *Thermopower in the correlated hopping regime*, Phys. Rev. B **13**, 647 (1976).

[129] K. Miyake and H. Kohno, *Theory of quasi-universal ratio of Seebeck coefficient to specific heat in zero-temperature limit in correlated metals*, J. Phys. Soc. Jpn. **74**, 254 (2005).

[130] V. Zlatic, B. Horvatic, I. Milat, B. Coqblin, G. Czycholl, and C. Grenzebach, *Thermoelectric power of cerium and ytterbium intermetallics*, Phys. Rev. B **68**, 104432 (2003).

[131] I. Milat, Ph.D. thesis, ETH Zurich (2006).

[132] W. Koshibae, K. Tsutsui, and S. Maekawa, *Thermopower in cobalt oxides*, Phys. Rev. B **62**, 6869 (2000).

# BIBLIOGRAPHY

[133] L. D. Hicks and M. S. Dresselhaus, *Effect of quantum-well structures on the thermoelectric figure of merit*, Phys. Rev. B **47**, 12727 (1993).

[134] H. Ohta, S. Kim, Y. Mune, T. Mizoguchi, K. Nomura, S. Ohta, T. Nomura, Y. Nakanishi, Y. Ikuhara, M. Hirano, et al., *Giant thermoelectric Seebeck coefficient of a two-dimensional electron gas in $SrTiO_3$*, Nat. Mater. **6**, 129 (2007).

[135] W. Kohn, *Theory of the Insulating State*, Phys. Rev. **133**, A171 (1964).

[136] A. J. Millis and S. N. Coppersmith, *Interaction and doping dependence of optical spectral weight of the two-dimensional Hubbard model*, Phys. Rev. B **42**, 10807 (1990).

[137] P. F. Maldague, *Optical spectrum of a Hubbard chain*, Phys. Rev. B **16**, 2427 (1977).

[138] B. S. Shastry and B. Sutherland, *Twisted boundary conditions and effective mass in Heisenberg-Ising and Hubbard rings*, Phys. Rev. Lett. **65**, 243 (1990).

[139] A. Comanac, L. de' Medici, M. Capone, and A. J. Millis, *Optical conductivity and the correlation strength of high-temperature copper-oxide superconductors*, Nat. Phys. **4**, 287 (2008).

[140] J. M. Ziman, *Electrons and Phonons* (Clarendon Press, Oxford, 1960).

[141] G. D. Mahan, *The best thermoelectric*, Proceedings of the National Academy of Sciences **93**, 7436 (1996).

[142] G. D. Mahan, *Many-particle physics* (Plenum Press, New York, 1981).

[143] N. W. Ashcroft and N. D. Mermin, *Solid State Physics* (Brooks Cole, 1976).

[144] G. Pàlsson and G. Kotliar, *Thermoelectric Response Near the Density Driven Mott Transition*, Phys. Rev. Lett. **80**, 4775 (1998).

[145] D. A. Broido and T. L. Reinecke, *Effect of superlattice structure on the thermoelectric figure of merit*, Phys. Rev. B **51**, 13797 (1995).

# BIBLIOGRAPHY

[146] W. Koshibae and S. Maekawa, *Effects of Spin and Orbital Degeneracy on the Thermopower of Strongly Correlated Systems*, Phys. Rev. Lett. **87**, 236603 (2001).

[147] S. Mukerjee and J. E. Moore, *Doping dependence of thermopower and thermoelectricity in strongly correlated materials*, Appl. Phys. Lett. **90**, 112107 (2007).

[148] S. D. Huber, E. Altman, H. P. Buchler, and G. Blatter, *Dynamical properties of ultracold bosons in an optical lattice*, Phys. Rev. B **75**, 085106 (2007).

[149] V. Zlatic, in *Lectures on the Physics of Strongly correlated Systems XII*, edited by A. Avella and F. Mancini (AIP, 2008), vol. 1014, pp. 198–247.

[150] C. A. Domenicali, *Irreversible Thermodynamics of Thermoelectricity*, Rev. Mod. Phys. **26**, 237 (1954).

[151] M. Jonson and G. D. Mahan, *Mott's formula for the thermopower and the Wiedemann-Franz law*, Phys. Rev. B **21**, 4223 (1980).

[152] J. K. Freericks, V. Zlatic, and A. M. Shvaika, *Electronic thermal transport in strongly correlated multilayered nanostructures*, Phys. Rev. B **75**, 035133 (2007).

[153] O. Betbeder-Matibet and P. Nozieres, *Transport equation for quasiparticles in a system of interacting Fermions colliding on dilute impurities*, Ann. Phys. (NY) **37**, 17 (1966).

# Acknowledgments

I would like to express my gratitude to my supervisor Manfred Sigrist. I would also like to thank my co-examiner Thilo Kopp for the critical reading of the manuscript.

Die VDM Verlagsservicegesellschaft sucht für wissenschaftliche Verlage abgeschlossene und herausragende

## Dissertationen, Habilitationen, Diplomarbeiten, Master Theses, Magisterarbeiten usw.

für die kostenlose Publikation als Fachbuch.

Sie verfügen über eine Arbeit, die hohen inhaltlichen und formalen Ansprüchen genügt, und haben Interesse an einer honorarvergüteten Publikation?

Dann senden Sie bitte erste Informationen über sich und Ihre Arbeit per Email an *info@vdm-vsg.de*.

**Sie erhalten kurzfristig unser Feedback!**

VDM Verlagsservicegesellschaft mbH
Dudweiler Landstr. 99              Telefon  +49 681 3720 174
D - 66123 Saarbrücken              Fax      +49 681 3720 1749
**www.vdm-vsg.de**

Die VDM Verlagsservicegesellschaft mbH vertritt

Printed by Books on Demand GmbH, Norderstedt / Germany